Methods of Argumentation

Argumentation, which can be abstractly defined as the interaction of different arguments for and against some conclusion, is an important skill to learn for everyday life, law, science, politics and business. The best way to learn it is to try it out on real instances of arguments found in everyday conversational exchanges and legal argumentation. The introductory chapter of this book gives a clear general idea of what the methods of argumentation are and how they work as tools that can be used to analyze arguments. Each subsequent chapter then applies these methods to a leading problem of argumentation. Today the field of computing has embraced argumentation as a paradigm for research in artificial intelligence and multi-agent systems. Another purpose of this book is to present and refine tools and techniques from computing as components of the methods that can be handily used by scholars in other fields.

Douglas Walton currently holds the Assumption Chair in Argumentation Studies and is Distinguished Research Fellow of the Centre for Research in Reasoning, Argumentation and Rhetoric at the University of Windsor, Canada. His most recent book is *Argumentation Schemes*, coauthored with Chris Reed and Fabrizio Macagno (Cambridge University Press, 2008). Walton's work has been used to better prepare legal arguments and to help develop artificial intelligence. His books have been translated worldwide, and he attracts students from many countries to study with him.

Methods of Argumentation

DOUGLAS WALTON

University of Windsor, Canada

CAMBRIDGE
UNIVERSITY PRESS

CAMBRIDGE
UNIVERSITY PRESS

University Printing House, Cambridge CB2 8BS, United Kingdom

One Liberty Plaza, 20th Floor, New York, NY 10006, USA

477 Williamstown Road, Port Melbourne, VIC 3207, Australia

314-321, 3rd Floor, Plot 3, Splendor Forum, Jasola District Centre, New Delhi - 110025, India

79 Anson Road, #06-04/06, Singapore 079906

Cambridge University Press is part of the University of Cambridge.

It furthers the University's mission by disseminating knowledge in the pursuit of education, learning and research at the highest international levels of excellence.

www.cambridge.org
Information on this title: www.cambridge.org/9781107677333

© Douglas Walton 2013

First published 2013

A catalogue record for this publication is available from the British Library

Library of Congress Cataloging in Publication data
Walton, Douglas N.
Methods of argumentation / Douglas Walton, University of Windsor, Canada.
pages   cm
Includes bibliographical references and index.
ISBN 978-1-107-03930-8 (hardback) – ISBN 978-1-107-67733-3 (paperback)
1. Reasoning.   I. Title.
BC177.W3247   2013
168–dc23   2013015608

ISBN 978-1-107-03930-8 Hardback
ISBN 978-1-107-67733-3 Paperback

For Karen, with love

Contents

Acknowledgments

I would like to acknowledge support for the work in this book from Social Sciences and Humanities Research Council of Canada Insight Grant 435-2012-0104. For comments received when material in some chapters of the book was presented at meetings of the Centre for Research in Reasoning, Argumentation and Rhetoric, I would like to thank J. Anthony Blair, Marcello Guarini, Hans V. Hansen, Catherine Hundelby, Ralph Johnson, Robert C. Pinto and Christopher W. Tindale. I would also like to thank Floris Bex, Thomas F. Gordon, Fabrizio Macagno and Henry Prakken for discussions that helped to refine formulations of my views in some chapters and correct some errors in them. An earlier version of Chapter 8 was read to the Department of Theoretical Philosophy at the University of Groningen on September 29, 2011. I would like to thank the following participants for comments and discussions: Barteld Kooi, Erik Krabbe, Henry Prakken, Allard Tamminga, Jan Albert van Laar, Rineke Verbrugge, Bart Verheij and Jean Wagemans.

Five of the chapters of this book are based on papers previously published in refereed journals. These five papers have been substantially revised, updated and edited. The editing work has included the addition of new material to these chapters, as well as deletion of some material and distribution of other material between one chapter and another. Permission to reprint each of them has been obtained from the copyright owner. Chapter 2 is based on the paper 'How to Refute an Argument Using Artificial Intelligence', originally published in *Studies in Logic, Grammar and Rhetoric*, 23(36), 2011, 123–154. Chapter 4 is based on the paper 'Argument Mining by Applying Argumentation Schemes', originally published in *Studies in Logic*, 4(1), 2011, 38–64. Chapter 5 is based on the paper 'Similarity, Precedent and Argument from Analogy', originally published in *Artificial Intelligence and Law*, 18 (3), 2010, 217–246. Chapter 6 is based on the paper 'Teleological Argumentation to and from Motives', originally published in *Law, Probability and Risk*, 10, 2011, 203–223. Chapter 8 is based on the paper 'Defeasible Reasoning and Informal Fallacies', originally published in *Synthese*, 179(3), 2011, 377–407. For help with indexing, proofreading and improving some of the figures, I would like to thank (respectively) Rita Campbell, Lynn Fraser and Mike Busuttil.

D.W.

1

Introducing Some Basic Concepts and Tools

Argumentation, which can be abstractly defined as the interaction of different arguments for and against some conclusion, is an important skill to learn for everyday life, law, science, politics and business. It is a rich, interdisciplinary area of research straddling philosophy, communication studies, linguistics, psychology and artificial intelligence that has developed context-sensitive practical methods to help a user identify, analyze and evaluate arguments.

Recently, the field of computing has embraced argumentation as a paradigm for research in artificial intelligence and multi-agent systems. Artificial intelligence in particular has seen a prolific growth in uses of argumentation. Argumentation has proved helpful to computing because it has provided concepts and methods used to build software tools for designing, implementing and analyzing sophisticated forms of reasoning and interaction among rational agents (Reed and Grasso, 2007). Recent successes include argumentation-based models of evidential relations and legal processes of examination and evaluation of evidence. Argument mapping has proved to be a useful tool for designing better products and services and for improving the quality of communication in social media by making deliberation dialogues more efficient. There now exist formal systems of argumentation to model many aspects of reasoning and argument that were formerly studied only by less structured methods of informal logic.

Now there has been the starting of a feedback loop. The formal argumentation methods and concepts that were developed in artificial intelligence are themselves being used to refine informal argumentation methods. In the past the argumentation methods have come from the humanities and social sciences. Now the tools that have been developed to model the features and problems of argumentation in natural language discourse, and other special contexts such as legal reasoning, are computationally precise. One benefit of this reverse transfer to and from computer science has been the refinement of argumentation theory itself through mathematically precise modeling of its core concepts and methods. The purpose of this book is to present and refine these tools from computing to help build a better set of methods that can be used by all argumentation scholars. This project is carried forward by showing how the new methods can already be applied with some success to some of the leading problems of argumentation.

The book gives a clear idea of what the methods are and how they work as tools that can be used to study arguments. To distinguish itself from other views of argumentation, the book calls its approach "logical argumentation", suggesting a joining of informal logic with the formal argumentation technology of computer science. Each chapter of the book applies these methods to a leading problem of argumentation studies. The problems studied in the book include the problems of defining the notions of critical questioning, undercutting, rebuttal and refutation, (2) the problem of representing critical questions on an argument diagram, (3) the pervasive problem of finding the missing premises (or conclusions) in an argument, (4) the problem of applying argumentation schemes to real arguments in natural language discourse, (5) the problem of modeling how argument from precedent in our system of law is based on a form of argument from analogy that can represent the notion of similarity between cases, (6) the problem of reasoning backward from external data to a hypothesis about an agent's presumed internal state of mind, for example, its motive, (7) the problem of understanding how scientific inquiry begins from a discovery phase, and (8) the problem of how an arguer's position can be adequately and fairly represented in order to properly criticize or refute it. The substantial progress made in the book on solving these problems demonstrates how useful the methods of logical argumentation can be.

1. Logical Argumentation as a Distinctive Theory

The roots of logical argumentation are in informal logic, a discipline that has the goal of providing criteria and methods to help students identify, analyze and evaluate arguments found in a natural language text of discourse. Logical argumentation originally came out of forty years of experience in teaching critical thinking skills to university students and research studying informal fallacies, and is now being widely used and tested in many fields. It was originally based on collecting many examples of arguments from everyday conversational discourse and from law, analyzing them, visualizing them and evaluating them as case studies, and solving common problems posed by features of the arguments in the case studies. As the simple argument maps presented below show, logical argumentation arises from the practical task of assisting users to analyze and critically evaluate arguments of the kind found in everyday conversational discourse and in other contexts such as legal and scientific argumentation.

However, logical argumentation as a theory is wider than the traditional focus of informal logic, because of its integration with artificial intelligence and because of its aim of providing assistance with the task of argument invention, as well as the tasks of argument identification, analysis and evaluation. Logical argumentation is a theory that can be applied to many fields, including informal logic, speech communication, artificial intelligence and linguistics. It is interdisciplinary, even though it has central affiliations with the field of informal logic, because it has connections in communication studies and is increasingly being applied in computer science, especially in artificial intelligence and multi-agent systems.

The author has written many books on specific topics in argumentation studies and informal logic, and also textbooks meant to explain to students how the methods built and refined in these books can be learned and applied to examples of arguments in everyday conversational discourse. However, there has been no single book that attempts to put all these results together in a unified approach to logical argumentation. This book binds the research results of the findings in computer science together by distilling out of them a distinctive theory underlying an accompanying set of methods for the identification, analysis and evaluation of arguments. It is easy to explain the theory in general outline. But to understand how the theory works, you need to see how it applies to real examples and how it is used to solve significant problems of argumentation. The chapters of this book show the theory being applied to a series of real examples and problems of argumentation. The chapters present examples representing specific instances that give rise to problems to which the theory is applied. The attempts to solve the problems show the methods at work, and cumulatively show how the methods give rise to a general theory that fits them together.

This theory has three broad characteristics as an approach to rational cognition. First, it is procedural, meaning that proving something is taken to be a sequence with a start point, an end point, and an interval in between representing a sequence of orderly steps. The second characteristic is that it does not aim to prove something is true as knowledge that must be accepted beyond all doubt, but recognizes the bounds of human rationality required by the need to make decisions under conditions of uncertainty and lack of knowledge. This second characteristic is called bounded rationality. The third characteristic is the viewing of intelligence as a social process that is not located exclusively in our individual brains. This characteristic holds that two heads are better than one, implying that even when a single agent reasons by deliberating about what to do or what claim to accept based on evidence, it does this best by examining the evidence on both sides, pro and contra. Thus whether one agent is involved or a group of agents is engaged in deciding what to do or to accept based on the evidence, rational thinking is best seen as a dialogue process in which arguments are put forward by one side and critically questioned by the other side.

2. The Methods and the Theory

Logical argumentation is a distinctive philosophical viewpoint built around a set of practical methods to help a user identify, analyze and evaluate arguments in specialized areas such as law and science, as well as arguments of the kind used in everyday conversational discourse. The method of logical argumentation has twelve defining characteristics.

1. The procedure for examining and criticizing the arguments on both sides forms a dialogue structure in which two sides take turns putting forward speech acts (e.g., making assertions or asking questions).
2. The dialogue has rules for incurring and retracting commitments that are activated by speech acts. For example, when a participant makes an assertion, he or she becomes committed to the proposition contained in the assertion.
3. The method uses the notion of commitment (or acceptance; Freeman, 2005) as the fundamental tool for the analysis and evaluation of argumentation rather than the notion of belief. The reason is that belief is a psychological notion internal to an agent (Walton, 2010a).
4. The method assumes a database of commonly accepted knowledge that, along with other commitments, provides premises for arguments. The knowledge base is set in place at the opening stage, but can be revised as new relevant information needs to be collected and considered.
5. The procedure is dynamic, meaning that it continually updates its database as new information comes in that is relevant to an argument being considered.
6. The arguments advanced are (for the most part) defeasible, meaning that they are subject to defeat as new relevant evidence comes in that refutes the argument.
7. Conclusions are accepted on a presumptive basis, meaning that in the absence of evidence sufficient to defeat it, a claim that is the conclusion of an argument can be tentatively accepted, even though it may be subject to later defeat as new knowledge comes in.
8. The method analyzes and evaluates argumentation concerning a claim where there is evidence for it as well as against it. Thus any argument is subject to critical questioning until closure of the dialogue.
9. The dialogue uses critical questioning as a way of testing plausible explanations and finding weak points in an argument that raise doubt concerning the acceptability of the argument.

10. The method uses standards of proof. Criteria for acceptance are held to depend on standards that require the removal of specifiable degrees of reasonable doubt.
11. The methods applied include defeasible argumentation schemes, deductive arguments, inductive arguments, presumptive arguments and argument visualization software tools.
12. The method comprises the study of explanations as well as arguments, including the form of argument called inference to the best explanation, or abductive reasoning.

There are two (often opposed) models of rational thinking and acting in the literature on cognitive science, and logical argumentation theory has a preference for one of these models as an approach to be taken in applying its methods, even though it acknowledges the need for both of them. The belief-desire-intention (BDI) model is based on the concept of an agent that carries out practical reasoning based on goals that represent its intentions and incoming perceptions that update its set of beliefs. According to the account of rational thinking of the BDI model, an agent has a set of beliefs that are constantly being updated by sensory input from its environment, and a set of desires (wants) that are then evaluated (by desirability and achievability) to form intentions. For example, on Bratman's (1987) version of the BDI model, forming an intention is described as part of adopting a plan that includes the agent's desires (wants) and beliefs.

According to the commitment model, agents interact with each other in a dialogue in which each contributes speech acts. Commitments are statements that the agent has expressed or formulated, either alone or as part of a group deliberation, and has pledged to carry out or has publicly asserted. Each agent has a commitment set, and as the one asks questions that the other answers, commitments are inserted into or retracted from each set, depending on the move, which takes the form of a speech act, that each speaker makes. A commitment is essentially a proposition that an agent has gone on record as accepting as indicated by a transcript or some other evidence that can be used to pin down exactly what the speaker said (Hamblin, 1970; 1971). One highly significant difference between the two models is that desires and beliefs are private psychological notions internal to an agent, while commitments are statements externally accepted by an agent and recorded in an external memory that is transparent to all parties.

The logical argumentation model and its accompanying set of methods take a view of proof and justification different from that taken in current epistemology in analytical philosophy, which is based on a true belief framework. On this approach, knowledge is taken to be true belief plus some third component, usually called justification. On the logical argumentation approach, knowledge is seen as a form of belief firmly fixed by an argumentation procedure that has examined the evidence on both sides, and uses standards of proof to conclude that the proposition in question can be proved. The justification of proof is that the evidence supporting the proposition is so much stronger than the evidence against it, or doubts that have been raised against it, that the proposition can be accepted as knowledge. However, on this evidentialist approach, knowledge, especially scientific knowledge, must be seen as defeasible. There are two especially important consequences of the view. One is that falsifiability is taken to be a criterion of genuine scientific knowledge. The other is that knowledge does not deductively imply truth.

To sum up, there are four main components of the methods used in logical argumentation theory: argumentation schemes, dialectical structure, argument mapping, and modeling in a computational argumentation system. These four components are described in the next four sections.

3. Argumentation Schemes

Logical argumentation is based on argumentation schemes, such as argument from expert opinion, that represent commonly used types of arguments that are defeasible. The schemes connect arguments together into sequences, often called chaining, by taking the conclusion of one argument as a premise in a subsequent argument. Schemes identify patterns of reasoning linking premises to a conclusion that can be challenged by raising critical questions. The names of some easy to recognize argumentation schemes are listed in Table 1.1.

A more complete list of twenty-nine such schemes will be given in Chapter 4. Some of these schemes appear to be subtypes of others. For example, argument from threat is a species of argument from negative consequences. It has the additional implication that the proponent is stating a readiness to carry out the negative consequences for the respondent.

As an easy example to appreciate, we can give the scheme for argument from expert opinion. It has two premises and a conclusion.

Major Premise: Source E is an expert in subject domain S containing proposition A.

Minor Premise: E asserts that proposition A is true (false).

Conclusion: A is true (false).

The following critical questions represent standard ways of casting the argument into doubt.

CQ$_1$: *Expertise Question.* How credible is E as an expert source?

CQ$_2$: *Field Question.* Is E an expert in the field that A is in?

CQ$_3$: *Opinion Question.* What did E assert that implies A?

CQ$_4$: *Trustworthiness Question.* Is E personally reliable as a source?

CQ$_5$: *Consistency Question.* Is A consistent with what other experts assert?

CQ$_6$: *Backup Evidence Question.* Is E's assertion based on evidence?

TABLE 1.1 *Some common argumentation schemes*

Argument from Witness Testimony	Argument from Verbal Classification	Argument from Rule
Argument from Expert Opinion	Argument from Appearances (Perception)	Argument from Threat
Argument from Analogy	Argument from Positive Consequences	Argument from Popular Opinion
Argument from Precedent	Argument from Negative Consequences	Direct *Ad Hominem* Argument (Personal Attack)
Practical Reasoning (Goal-Directed Reasoning to Act)	Circumstantial *Ad Hominem* Argument	Argument from Correlation to Cause
Argument from Evidence to a Hypothesis	Abductive Reasoning	Argument from Commitment
Argument from Ignorance (Negative Evidence)	Argument from Sunk Costs	Slippery Slope Argument

This form of argument is defeasible. If an expert says that a proposition is true, there may be a good reason for accepting it is true, but there may also be good reason for doubting whether it is true once it is pointed out that the expert is biased, for example, by evidence showing that he or she will gain financially from his or her claim. Defeasibility of arguments is very important in the logical argumentation model. The ideal arguer retracts his or her claim if it can be shown to be insufficiently supported by evidence that meets the appropriate standard of proof for accepting it. An ideal arguer is not only one who backs up his or her claims by supporting evidence, but also one who is open-minded. The ideal arguer probes into the reasons behind and those of his or her speech partner, formulating criticisms of his or her arguments. How this process works can be illustrated briefly with some examples.

4. Dialectical Structure

As noted in Section 2, logical argumentation reaches a decision on whether or not to accept a claim based on the arguments both for and against the claim; therefore on the logical argumentation point of view, an argument always has two sides, the pro and contra. They take turns making moves that contain speech acts. Some of the most common speech acts are identified in Table 1.2.

Speech acts are put forward by each participant at each move in the dialogue, and the structure of the dialogue is defined by rules (protocols) that set preconditions and post-conditions for the speech acts used in that type of dialogue.

TABLE 1.2 *Some common types of speech acts*

Speech Act	Dialogue Form	Function
Question (yes-no type)	S?	Speaker asks whether S is the case
Assertion (claim)	Assert S	Speaker asserts that S is the case
Concession (acceptance)	Accept S	Speaker incurs commitment to S
Retraction (withdrawal)	No commitment S	Speaker removes commitment to S
Challenge (demand for proof of claim)	Why S?	Speaker requests that hearer give an argument to support S
Put Argument Forward	$P_1, P_2, \ldots, P_n$ therefore S	$P_1, P_2, \ldots, P_n$ is a set of premises that give a reason to support S

TABLE 1.3 *Example of a profile of dialogue*

Move	Proponent	Respondent
1.	Video games do not lead to violence	Why do you think so?
2.	Dr. Smith says so, and he is an expert	Do you think he could be biased?
3.	What evidence do you have for saying that?	His research is funded by the video game industry
4.	What evidence do you have for saying that?	It was shown by a 2001 investigation of the Parents' Defense League

In the small example dialogue shown in Table 1.3, the proponent begins at move 1 by making a claim. The respondent then puts forward a challenge demanding proof for this claim. At move 2, the proponent takes his turn by putting forward an argument from expert opinion. The respondent then asks a critical question. The proponent then asks for evidence to support the question, and the respondent offers some. However, the proponent asks for further evidence and the respondent offers it.

This small dialogue represents what is called a profile of dialogue, a short sequence of moves that could be part of a much longer sequence of argumentation. These small examples of dialogues need to be put into a wider perspective, viewed as a dialectical process by Freeman (1991, xiii): "We see arguments generated through a challenge-response dialogue where the proponent of some thesis answers critical questions posed by a challenger". So viewed, the structure of arguments takes the form of a procedure that has a start point and an end point.

A dialogue is defined as a 3-tuple {O, A, C} where O is the opening stage, A is the argumentation stage, and C is the closing stage. Dialogue rules define what types of moves are allowed. At the opening stage, the participants

TABLE 1.4 *Seven types of dialogue*

Type of Dialogue	Initial Situation	Participant's Goal	Goal of Dialogue
Persuasion	Conflict of Opinions	Persuade Other Party	Resolve Issue
Inquiry	Need to Have Proof	Verify Evidence	Prove Hypothesis
Discovery	Need an Explanation	Find a Hypothesis	Support Hypothesis
Negotiation	Conflict of Interests	Get What You Want	Settle Issue
Information	Need Information	Acquire Information	Exchange Information
Deliberation	Practical Choice	Fit Goals and Actions	Decide What to Do
Eristic	Personal Conflict	Hit Out at Opponent	Reveal Deep Conflict

agree to take part in some type of dialogue that has a collective goal. Each party has an individual goal and the dialogue itself has a collective goal. The initial situation is framed at the opening stage, and the dialogue moves through the opening stage toward the closing stage. In Table 1.4, the type of dialogue is identified in the left column and its main properties are identified in the three matching columns on the right.

In a persuasion dialogue the proponent has a thesis to be proved, his ultimate *probandum*, and the respondent can either have (1) the role of casting doubt on the proponent's attempts to prove his thesis or (2) the role of arguing for the opposite thesis. A rational arguer is one who follows the protocols for the type of dialogue appropriate for the argumentation in which he is engaged and whose arguments conform to the requirements of argumentation schemes. The goal of a persuasion dialogue is to reveal the strongest arguments on both sides by pitting one against the other to resolve the initial conflict posed at the opening stage. Each side tries to carry out its task of proving its ultimate thesis to the standard required to produce an argument stronger than the one produced by the other side. This burden of persuasion is set at the opening stage. Meeting one's burden of persuasion is determined by coming up with a strong enough argument using a chain of argumentation in which individual arguments in the chain are of the proper sort. To say that they are of the proper sort means that they fit argumentation schemes appropriate for the dialogue. 'Winning' means producing an argument that is strong enough to discharge the burden of persuasion set at the opening stage.

In deliberation dialogues decisions are also made, but the starting point of the dialogue is an issue about which action to take to achieve some goal, not an issue about whether a proposition is true or false. The party who raises the issue does not have a burden of persuasion. Indeed, even once positions (proposals for resolving the issue about which action to take) have been put forward in a deliberation dialogue, the parties who put forward the positions do not necessarily have a burden of persuasion. The proposals may have been put forward during a brainstorming phase of the deliberation, and a party may actually prefer some proposal put forward by some other party, after arguments about the pros and cons have been exchanged.

During the same sequence of argumentation an argument may shift from one type of dialogue to another. In some cases the shifts are based on an underlying embedding from the one dialogue into the other. In that case, the move to the second dialogue can support the chain of argumentation coming from the first type of dialogue. This kind of shift can be a good thing, from an argumentation point of view. For example, in a deliberation, the argumentation may shift to an information-seeking phase where facts relevant to the deliberation are brought into play. In other cases, the shift to the second dialogue can block the argumentation in the first dialogue, in some instances leading to fallacies. The analysis and evaluation of arguments in dialogues is based on procedural rules for the dialogue, as well as notions such as burden of proof and standard of proof, that set requirements for how strong an argument needs to be in order to be judged successful.

The logical argumentation model is normative because it sets standards for logical inference based on argumentation schemes and procedural standards that give requirements for how to take part in a dialogue with a speech partner. These standards can be structured in formal models of dialogue. However, the model is also partly empirical in that its purpose is to study real arguments used in everyday conversational discourse and other special contexts such as legal and scientific reasoning. For this reason, the logical argumentation model is based on the study of real examples of arguments. As well as identifying, analyzing and evaluating arguments used in a given case, logical argumentation also has the capability for constructing arguments. This technology is based on the application of argumentation schemes, with their capability to represent implicit parts of the text, especially implicit premises and conclusions in a chain of reasoning. The given arguments in a knowledge base can be chained forward toward the ultimate conclusion to be proved.

5. Rationale and Araucaria

In its simplest form, an argument diagram, or argument map as it is equivalently called, is composed of two elements: a set of propositions representing premises or conclusions of arguments and a set of arrows representing inferences from some propositions to others. For this reason an argument map is often called a box and arrow diagram, a visual representation of an argument formed by drawing arrows leading text boxes to other text boxes. An argument diagram takes the form of a tree structure in which there is a single proposition representing the ultimate claim or thesis to be proved at the root of the tree. All the other propositions are premises or conclusions that lead along branches of the tree to this root proposition.

An argument diagram can easily be made using pencil and paper, but nowadays there are many argument visualization tools that can be used to assist in creating an argument diagram that can be saved and later modified. Such argument mapping tools have now become centrally important logical argumentation methods in their own right, as they can perform different functions that are helpful for clarifying, analyzing, summarizing and evaluating arguments. Below some simple examples are given to show how these methods work and to show some key differences between different visualization tools. Those that are conceptually interesting will be studied further in Chapter 2. There are now more than sixty software systems for argument visualization (Scheuer et al., 2010) that can be used to summarize or analyze argumentation in a visual format on a computer screen for various different kinds of purposes. However, of the available systems, we choose only four to introduce here. These four are easy to use and have other features that make them attractive for several reasons from a point of view of argumentation studies. Also, each of them has certain fundamental features that enable significant contrast to be drawn between them that will turn out to be important for our purposes in dealing with some fundamental problems of argumentation, especially in Chapter 2.

Rationale (http://rationale.austhink.com/) is a software tool for working with argument maps to help students get a better grasp of good essay writing structure, to learn skills of thinking critically and to prepare for debates. Of the four systems, it is perhaps the easiest for the beginner to start using. A Rationale argument map is drawn in the form of a tree structure with the contention, the main issue or topic under consideration, represented in a text box at the top of the page. The premises and conclusions of the argument supporting or attacking the main contention are placed in text boxes that lead to the main contention by a series of arrows drawn as lines. A reason is a type of premise and an argument that directly supports a contention, while an objection represents that which directly refutes a contention. So a reason represents positive support of the claim, while an objection represents a negative type of argument that attacks or undermines a claim. Rationale distinguishes between two kinds of negative arguments of this sort called objection and rebuttal, but we will defer the discussion of this distinction to Chapter 2.

To give a simple example of how a Rationale argument map looks, Figure 1.1 is an argument map of an example drawn using Rationale. The example is a case of argument from expert opinion where one expert makes a certain claim while another expert makes the opposite claim. One expert, Dr. Smith, makes the claim that video games do not lead to violence. But the other expert, Dr. Jones, states the opposite proposition, namely, that video games do lead to violence. This example is a classic case of what is often called the battle of the experts.

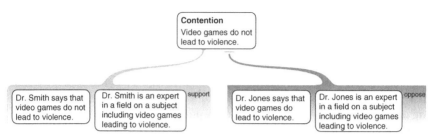

FIGURE 1.1 Rationale Argument Map of the Video Games Example

As shown in Figure 1.1, Rationale visualizes the argument from Dr. Smith on the left, which is a supporting argument presenting evidence in favor of the conclusion that video games do not lead to violence. The argument on the left is an example of a linked argument, meaning that the two premises go together to support the conclusion. Rationale can also be used to classify the argument as an instance of the argumentation scheme for argument from expert opinion by enabling the user to label the right premise – the statement that Dr. Smith is an expert in a field on the subject, including video games leading to violence – as an instance of an expert opinion. But we want to emphasize here is the argument on the right in Figure 1.1, an argument separate from the supporting argument used in opposition to it. This opposing argument is represented as an argument separate from expert opinion that attacks the contention that video games do not lead to violence.

It will be interesting to show the reader at this point how different argument mapping tools can represent such a relation of opposition in a different pictorial manner. Araucaria is a software tool for analyzing arguments that helps a user to reconstruct and diagram a given argument using a point-and-click interface.[1] Araucaria supports argumentation schemes and provides a user-customizable set of schemes that can also be helpful when analyzing arguments.

To analyze a given argument, the user begins with a text document containing an argument that has been cut and pasted from its source, for example, a newspaper article. Then he or she chooses each of the propositions that function as premises or conclusions of the argument by highlighting them from the list on the left side of the screen, and transfers them to a box that appears on the right side of the screen. The user then draws in the arrows displaying the inferences from the premises to the conclusions. The user can also label each argument with an argumentation scheme by calling down a menu that presents lists of argumentation schemes.

This menu for the video games example is shown in Figure 1.2.

---

[1] Araucaria is freeware built by Chris Reed and Glenn Rowe. It can be downloaded from this site: http://araucaria.computing.dundee.ac.uk/.

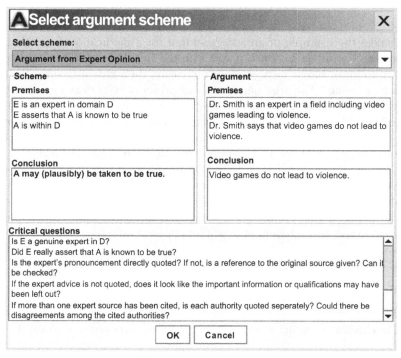

FIGURE 1.2 Menu Applying Scheme to Argument

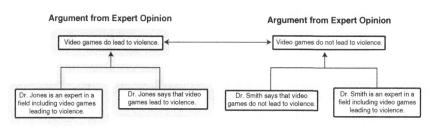

FIGURE 1.3 Araucaria Argument Map of the Video Games Example

Once the user has applied one or more argumentation schemes to selected arguments in the argument diagram in this fashion, Araucaria is used to build an argument diagram of the kind displayed in Figure 1.3. Note that in Figure 1.2 one of the premises is missing in the argument. This observation reveals how schemes can be used to find missing premises.

Araucaria has some other distinctive features concerning the representation of argument rebuttal, which will be illustrated in the next example.

6. An Example of Refutation

Let's take the example of an advertisement for a medication for diabetes (*Newseek*, November 26, 2007, p. 25) with the headline: "ACTOS has been shown to lower blood sugar without increasing their risk of having a heart attack or stroke". The argument in this ad presents ACTOS as a way for the reader who has type 2 diabetes to solve the problem of lowering his or her blood sugar by using practical reasoning. It says: you have the goal of lowering your blood sugar; taking ACTOS is a means to realize this goal; therefore you should take ACTOS. The argument fits the scheme for practical reasoning. When you put forward an argument based on practical reasoning, you are arguing to your respondent as follows: you have a goal, or want to solve a problem; this action I am proposing will help you to attain that goal, or will solve the problem; therefore you should carry out this action. This simplest form of practical reasoning, often called practical inference, is given in the following argumentation scheme (Walton, Reed and Macagno, 2008, 323). This scheme will be studied in more depth in Chapter 4, Section 3.

Major Premise: I have a goal G.

Minor Premise: Carrying out this action A is a means to realize G.

Conclusion: I ought (practically speaking) to carry out this action A.

Many arguments for health products fit this scheme.

Once you have identified the scheme, you can also identify some critical questions to think about. The set of critical questions matching the scheme for practical inference is from the account in (Walton, Reed and Macagno, 2008, 323).

CQ$_1$: What other goals do I have that should be considered that might conflict with G?

CQ$_2$: What alternative actions to my bringing about A that would also bring about G should be considered?

CQ$_3$: Among these alternative actions to bringing about A, which is arguably the most efficient?

CQ$_4$: What grounds are there for arguing that it is practically possible for me to bring about A?

CQ$_5$: What consequences of my bringing about A should also be taken into account?

The last critical question, CQ$_5$, sometimes called the side effects question, concerns potential negative consequences of the action. One of the side effects of taking this particular medication might be to increase the risk of having a heart attack or stroke. To anticipate the possibility of this critical question being raised, the advertisement explicitly states that the medication can lower blood sugar without the risk of heart attack or stroke. The

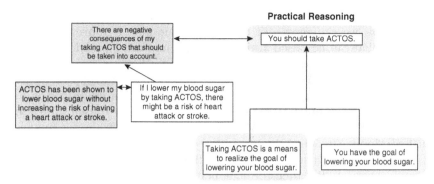

FIGURE 1.4 Argument in the ACTOS Example Visualized in Araucaria

argument in this advertisement is an example of proleptic argumentation, which can be defined as the anticipation and answering of an objection or opposed argument before one's opponent has actually put it forward. Now let's see how to represent this argument diagrammatically.

The argumentation in the ACTOS example may be represented in Araucaria as shown in Figure 1.4. Each premise or conclusion in the chain of argumentation is represented as a statement inside a text box. An inference from one statement to another is represented as an arrow. The argument on the right is shown as a linked argument, where the two premises function together to support the conclusion. The application of the argumentation scheme for practical reasoning is displayed by the shaded outline on the arrow leading from its premises to its conclusion.

The statement that there are negative consequences of my taking ACTOS that should be taken into account is represented as a refutation of the statement that one should take ACTOS. Refutation is drawn as a double-headed arrow, and the statement that is meant to be the refutation appears in a darkened box. Refutation is meant to be like negation in Araucaria. The statement 'If I lower my blood sugar by taking ACTOS, there might be a risk of heart attack or stroke' is shown as the reason supporting the statement that appears above it in the darkened box. Finally, there is another refutation shown in Figure 1.4. The statement that ACTOS has been shown to lower blood sugar without increasing the risk of having a heart attack or stroke is shown as a refutation of the statement that appears in the box without a darkened background that appears to the right of it. Just below that box, the statement on the left, also shown in a darkened box, is represented as a refutation of the statement that appears just the right of it. The double arrow in this instance is very small but is meant to represent a refutation. In effect, the structure of the argumentation shown in Figure 1.4 is that of a refutation of a refutation.

The problem posed by this example is that of representing critical questions on an argument diagram. We want to represent the critical question 'Are there negative consequences of my taking ACTOS?' on the diagram.

However, we can insert statements only in the text boxes. The problem then is how we can represent this critical question as representing an objection that can be raised against the argument using practical reasoning represented on the right side of Figure 1.4. We also want some way of representing on the diagram how the statement used to represent the critical question in Figure 1.4 functions as an objection or refutation to the argument from practical reasoning. Is this refutation by itself sufficient to defeat the argument based on practical reasoning? Or does it require further evidence to defeat the argument? In the argument diagram shown in Figure 1.4 another statement is given that supports the negative consequences assertion. But then the statement itself is objected to by another refutation.

7. The ArguMed System

Verheij (2005) constructed an argument diagramming method called ArguMed to represent argumentation schemes and to show how they apply to arguments in a given case. In ArguMed, blocking moves that make an argument default are drawn by a device called entanglement. Entanglement is represented as a line that meets another line at a junction marked by an X, indicating that new evidence attacks the inferential link between the premises and conclusion of the original argument, making the original argument default. Figure 1.5 shows how entanglement can be represented by ArguMed.

The idea of entanglement is that the connection between a reason and its conclusion, represented by the arrow between them, can be subject to doubt, just like other claims. Such argumentation can be either supporting, by giving reasons for the connection between the reason and its conclusion, or rebutting, by giving reasons against the connection between the reason and its conclusion. The positive version of entanglement is represented graphically by an ordinary arrow pointing to another arrow. The negative version of entanglement can be represented graphically by an arrow ending in a circled X pointing to another arrow.

In Figure 1.5, the method of diagramming is similar to that of Araucaria in that the premises and the conclusions in the chain of argumentation appear as statements in text boxes, and the inferences from a set of premises to a conclusion are represented as arrows. Both ArguMed and Araucaria can represent the distinction between linked and convergent arguments. The inference shown at the top of Figure 1.5 has its two premises joined by practical reasoning, making it linked. When the argument is shown in this way, it shows how the critical question represented in the box in the middle at the right undercuts the main argument to the conclusion that one should take ACTOS. There is a line going from this critical question to the arrow joining the premises to the conclusion of the prior inference, and the arrowhead is drawn as a circled X.

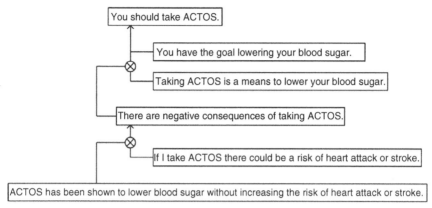

FIGURE 1.5 The Argumentation in the ACTOS Example Represented in ArguMed

An interesting thing about the diagram shown in Figure 1.5 is that it shows a refutation of a refutation. The statement shown in the text box at the bottom, 'ACTOS has been shown to lower blood sugar without increasing the risk of having a heart attack or stroke' is shown as a refutation of the critical question shown in the middle box on the right that itself functions as a refutation of the prior argument based on practical reasoning to the conclusion that you should take ACTOS. ArguMed is different from Araucaria in that it enables the representation of entanglement. Its distinctive feature is that it represents a statement meant as a refutation as an attack on the inference it was directed against, as contrasted with an attack the premises or the conclusion of the argument.

8. The Carneades Argumentation System

The Carneades Argumentation System (Gordon and Walton, 2009), named after the Greek skeptical philosopher Carneades, is a computational model of argumentation. Carneades is also a mathematical model consisting of definitions of mathematical structures and functions on these structures (Gordon, Prakken and Walton, 2007). Carneades has been implemented using a functional programming language, and has a graphical user interface used to draw argument diagrams (https://github.com/carneades/carneades). It also has a software library for building applications supporting other argumentation tasks. The version that presently exists can be used to analyze construct and evaluate arguments using defeasible forms of argument such as argument from testimony, argument from analogy, and many other kinds of arguments (Gordon, 2010).

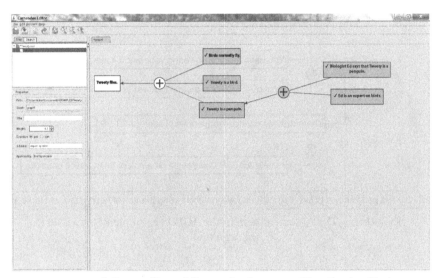

FIGURE 1.6 Screen Shot of Carneades Argument Visualization Tool

The actual example displayed in Figure 1.6 is not so important. It merely shows the Tweety example with the exception to the rule shown in the bottom box (Tweety is a penguin). Also shown is how Carneades evaluates argumentation. Although all the premises are accepted, indicated by the darkened text boxes and the check marks in them, the conclusion is not accepted. The reason is that exception makes the argument default.

How the Carneades Argumentation System represents argumentation will be studied in detail in Chapter 2, but even at this point it will be interesting for us to see how it visualizes arguments in a manner different from the three preceding systems. To begin with, in Carneades, the ultimate proposition is shown at left and the sequence of argumentation fans out to the right. In Rationale, as we saw, the ultimate conclusion in a sequence of arguments is always shown as a text box at the top of the diagram so that the argumentation supporting or attacking the central claim propagates downward from this root. The tree is upside down, so to speak.

In the video games example, the ultimate conclusion to be proved – the statement that video games do not lead to violence – is shown by Carneades at the left in Figure 1.7.

Instead of arrows leading directly from the premises to the conclusion, we have argument nodes in which the name of the argumentation scheme is inserted. Lines from the pair of premises in the linked argument go to the argument node, and an arrow then goes from the node to the conclusion of the argument. Figure 1.7 shows an argument with the ultimate conclusion that video games do not lead to violence, and shows two different arguments from expert opinion supporting that conclusion. The one at the top is a

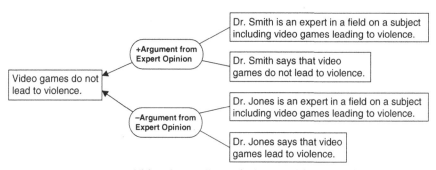

FIGURE 1.7 Video Games Example Drawn with Carneades

pro argument supporting the conclusion that video games do not lead to violence. The one at the bottom is a con argument attacking the conclusion that video games do not lead to violence. Both arguments fit the scheme for argument from expert opinion, and this information is shown in Figure 1.7, wherein each of the nodes has the name of the scheme attached to it. The scheme for argument from expert opinion is displayed in the node with the plus sign in it, representing a pro argument where the premises support the conclusion. There is another argument displayed in Figure 1.7 as well, an argument that is against the conclusion that video games do not lead to violence. It is also shown as an argument from expert opinion, but the minus sign in the node indicates that it is a contra argument.

The argument diagram in Figure 1.7 shows in general outline how arguments are mapped using the Carneades Argumentation System. The ultimate conclusion of the argument appears in the text box at the left of the diagram, and each argument is represented by a node that contains a plus or minus inside the node. The plus represents a pro argument and the minus represents a con argument. Each premise is represented by a proposition in a text box. An argument can have a single premise or multiple premises. When an argument is drawn with multiple premises leading to its node, it means that all the premises go together to support the conclusion. This type of argument is called a linked argument in informal logic. When two or more premises each independently support the conclusion, that structure is called a convergent argument in informal logic. In the Carneades Argumentation System, this type of argument is drawn as two or more separate arguments. In other words, it will have two or three nodes, depending on the number of independent premises supporting the conclusion. Information about the argumentation scheme for a particular argument is represented in the node of that argument.

Some differences between the Araucaria and Carneades styles of visualizing an argument can be summed up as follows. In an Araucaria diagram the ultimate conclusion is shown at the top, whereas in a Carneades diagram, the ultimate conclusion is shown at the left. The pro argument is represented

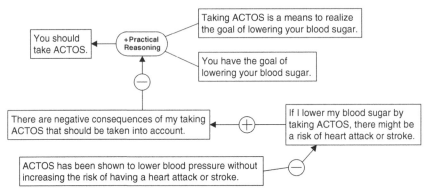

FIGURE 1.8 Carneades Version of the Argument Diagram for the Actos Example

by a plus sign in the node. The con argument is represented by a minus sign in the node. In addition to representing the premise-conclusion structure of an argument and the argumentation schemes joining a set of premises to a conclusion, Carneades can also be used to evaluate a pair of competing pro and con arguments such as those shown in Figure 1.5 to determine which of the pair can be evaluated as the argument to be accepted, based on standards of proof.

To see some other ways in which Carneades and Araucaria represent argumentation differently, let's turn back to the Actos example visualized using Araucaria in Figure 1.4 and using ArguMed in Figure 1.5. Turning to Figure 1.8, we see that the negative consequences critical question matching scheme for practical reasoning is represented by using a contra argument drawn by an arrow leading to the node containing the argumentation scheme for practical reasoning. In other words, Carneades structures the argumentation in this example by using entanglement, in contrast to the way Araucaria visualizes it in Figure 1.4.

We should also note that, the way the argument is visualized in Figure 1.8, there are negative consequences of my taking ACTOS that should be taken into account is backed up by supporting evidence, that if I lower my blood sugar by taking ACTOS, there might be a risk of heart attack or stroke. This statement is in turn backed up with further evidence by the statement at the bottom of Figure 1.8 stating that ACTOS has been shown to lower blood pressure without increasing the risk of heart attack or stroke. The contrast with how Araucaria diagrams the same argument raises some serious questions about how to analyze the notions of critical question, rebuttal and refutation using argument mapping tools. The statement that there are negative consequences of my taking ACTOS should be taken into account as raising the critical question, namely, the side effects question corresponding to the scheme for practical reasoning. So there are questions here concerning the precise relationship between cases where a critical question is used to cast doubt on argument and cases where a counterargument has been used to attack another argument, or even defeat it. Answering these questions will be the subject matter of Chapter 2.

9. Conclusions

During the first half of this chapter, logical argumentation was presented as a general theory with twelve leading characteristics that define it as a theoretical approach. But in the second half of chapter the introduction of argument tools such as argument mapping and argumentation schemes brought the theory down to earth because these tools are used to identify, analyze and evaluate real arguments in natural language texts of discourse. Through some simple examples, it was shown how the tools link with the theory to provide a system that has extensive capabilities for carrying out tasks that are highly significant. The fact that these tools, and the theory itself, have now been embedded in artificial intelligence models and software tools that can help the user to carry out these tasks of argument identification, analysis and evaluation makes logical argumentation a much more amenable area for study and academic research.

As illustrated in this chapter, one of the most useful developments in the evolution of the study of argumentation was the advent of software tools that aid an argument analyst to carry out the task of argument mapping (Reed, Walton and Macagno, 2007). Indeed, there are many more resources now becoming available for the study of argumentation through recent research in artificial intelligence. Just a few of the many subjects currently being investigated in artificial intelligence include multi-agent negotiation systems, argumentation-based models of evidential structures in legal reasoning, decision support systems using argumentation, and models of knowledge engineering. We also have automated argumentation systems used in computer-assisted collaborative learning, and single-user systems designed with the purpose of allowing a user to visualize, analyze and evaluate an argument in a given text (Reed and Grasso, 2007). Carneades artificial intelligence argumentation systems are especially well adapted to the needs of the logical argumentation theory.

Argumentation schemes in the Carneades model can be used as heuristic search procedures that apply statements from a database to find arguments pro or con the claim at issue. The arguments that turn up in the resulting stream are alternative ways that can be used to prove the claim. Requirements can be used to guide the search for arguments with acceptable premises. Rhetoric is concerned with presentation of arguments, and is a practical art directed toward helping develop the skill of persuading a particular audience to come to accept a particular claim or take a particular course of action. There is a close connection between logical argumentation and rhetoric, because generally audiences tend to be persuaded by arguments they think are reasonable. Therefore there is often a close approximation between an argument that is rhetorically persuasive and one that is reasonable from a point of view of logical argumentation. Also, the invention of new arguments has traditionally been held to be an important task for rhetoric. Logical argumentation can help the work of rhetorical scholars and practitioners by providing automated tools for argument construction.

An important aspect of logical argumentation was its use of a dialogue model so that argumentation is seen as a procedural sequence with a start point and an end point. During the argumentation stage of a dialogue, each side puts forward arguments that can be used to prove its central claim. They can do this by constructing chains of arguments designed to move forward through the argumentation stage to prove or disprove the central claim at issue. Each side makes attempts to prove its claim by taking the facts of the case as premises, along with other statements that are acceptable to the audience, and uses argumentation schemes to build a chain of argumentation that will go from these acceptable premises to the ultimate claim at issue. A successful chain of argumentation goes from the accepted premises as start points to the ultimate conclusion as the end point.

Logical argumentation is universal because the argumentation schemes and formal dialectical structures can be specified in an abstract and general way. Logical argumentation is also computationally implementable. However, to the extent that the model is applied to natural discourse, individual applications of it to different languages and fields are not universal. It is not known yet whether logical argumentation is able to capture all potential debates and arguments. There is no way yet to prove or disprove this claim. However, at its present development, it is widely applicable to many kinds of arguments and many settings where arguments are used.

10. Problems Investigated Next

The previous nine sections in this chapter have now introduced and explained some of the basic tools and concepts that can be applied to actual cases of arguments in order to identify, analyze and evaluate them. Argumentation is a field that anyone can engage in with a little training. The best way to learn it is to try it out on real instances of arguments found in texts of discourse, such as everyday conversational reasoning and legal reasoning. That is precisely what we will do in the next eight chapters. Each chapter picks out a serious and interesting problem that is at the forefront of argumentation studies and applies the tools explained in Chapter 1 to them.

There is a family of terms in argumentation that are closely related to each other and that all refer to some way in which a given argument is attacked, rebutted, refuted, undercut, critically questioned or objected to, thereby defeating it or casting it into doubt. The notions of critically questioning an argument and refuting an argument are perhaps the most fundamental concepts of logical argumentation. Proper understanding of this family of terms is fundamental to argumentation theory and to building argumentation technologies in artificial intelligence. Chapter 2 refines, clarifies and classifies them, using the Carneades Argumentation System. It begins with a simple example that illustrates two main ways of refuting an argument, and concludes with a seven-step procedure for seeking a refutation or objection.

Most arguments, when presented in a natural language text of discourse, cannot be properly understood, analyzed or evaluated without taking into account parts of the argument that were not explicitly stated in the text but are needed to properly make sense of the argument. These missing parts can be a premise, or some premises in the argument, or even the conclusion. Arguments with missing parts have traditionally been called enthymemes in the literature on logical argumentation at least since the Middle Ages, even though, as Chapter 3 will show, this terminology may be based on a misnomer. In Chapter 3, the traditional problem of enthymemes, the problem of finding the implicit premises and conclusions of an argument, is reconfigured by developing a comprehensive method of argument analysis. The comprehensive method employs the existing argumentation tools and concepts introduced in Chapter 1, which will be refined as they are applied in other parts of the book. These include argumentation schemes, argument mapping technology, common knowledge of the kind developed in artificial intelligence, and the notion of an arguer's set of commitments. How the comprehensive method works is illustrated with the use of two examples where missing parts of the argument are found and made explicit in an analysis represented on an argument map. As part of the comprehensive method, a set of requirements for identifying an argument in a text of discourse is developed that takes both the reasoning core and the dialectical level of an argument into account.

Chapter 4 investigates theoretical and practical aspects of applying argumentation schemes to real arguments and draws on details of how schemes have been modeled in argumentation systems. Examples of arguments are analyzed using the argumentation schemes for practical reasoning, the sunk costs argument, the slippery slope argument and arguments from consequences. Each of these types of arguments has been classified under the heading of informal fallacies in the logic textbooks, but it will be shown how each of them represents a reasonable but defeasible type of argument that holds generally but can fail in exceptional cases. Another interesting topic studied in Chapter 4 is the technique called argument mining, the systematic attempt to scan over the text of discourse and identify a specific type of argument occurring in it by using argumentation schemes. Chapter 4 also touches on the question of whether argumentation schemes can be classified into clusters where one scheme is closely related to others. The problem of classifying argumentation schemes is comparable to the biologists' problem of classifying plants or animals into species and subspecies.

One of the most encouraging developments in argumentation studies is the growing body of work that applies argumentation methods to legal reasoning. Argumentation theorists can find many good examples here to test out their theories. In Chapter 5, a famous case about the ownership of a valuable baseball that was hit into the stands is analyzed. Chapter 5 uses this example and others to show how the most central kind of case-based legal reasoning in our common law system is based on argumentation schemes employed in logical argumentation. It is shown (1) that there are two schemes for argument from analogy that seem to be competitors but are not, (2) how one of them is based on a distinctive type of similarity premise, (3) how to analyze the notion of similarity using story schemes illustrated by some cases, (4) how arguments from precedent are based on arguments from analogy, and in many instances arguments from classification, and (5) that when similarity is defined by means of story schemes, we can get a clearer idea of how it integrates with the use of argument from classification and argument from precedent in case-based reasoning by using a dialogue structure.

Chapter 6 is also about legal argumentation. Understanding how to model arguments that proceed from factual evidence about human actions to a hypothesis about the motive or intention that led to the action is a central problem not only for argumentation studies, but also for law and other fields such as history. Chapter 6 uses tools from argumentation and artificial intelligence to build a system to analyze reasoning from a motive to an action and reasoning from circumstantial evidence of actions to a motive. The tools include argument mapping, argumentation schemes, inference to the best explanation, and a hybrid method of combining argument and explanation. Several examples of use of relevant motive evidence in law are studied to illustrate how the system works. It is shown how adjudicating cases where motive of evidence is relevant depends on a balance of argumentation that can be tilted to one side or the other using plausible reasoning that combines arguments and explanations.

A new frontier for argumentation studies is the application of argumentation methods to scientific reasoning. Especially important in this connection is the problem of how scientific evidence used to support a hypothesis can be modeled using argumentation tools. In Chapter 7, the Carneades Argumentation System is used to model an example of the progress of a scientific inquiry starting from a discovery phase. During the discovery phase, (1) data are collected and used to construct hypotheses, (2) the hypotheses are tested by experiments and criticized, and (3) if one of them is strongly enough supported by the evidence to the required proof standard, it is tentatively accepted. That does not mean the statement asserted by the hypothesis is proved, however. For that to be done, the argumentation has to shift to a subsequent phase. The inquiry phase has the goal of proving the hypothesis to a suitable proof standard, disproving it, or proving that it cannot be proved or disproved. Chapter 7 reconfigures this problem as a dialectical one that requires a shift in the context of the argumentation from a discovery phase to an inquiry phase where the evidence both for and against hypothesis is marshaled.

The problem of analyzing informal fallacies, significant errors and systematic deceptions that represent classic cases where rational argumentation is going wrong provides a benchmark that can be used to test the worth of any serious theory of argumentation. There is a growing literature in argumentation studies on fallacies, but it is a central problem that so far there has been no widely accepted theory that enables us to give a general explanation of what a fallacy is. Chapter 8 puts forward a dialectical theory that argues that at least some of the main traditional fallacies should be considered as reasonable arguments when used as part of a properly conducted dialogue. It is shown that argumentation schemes, formal dialogue models and profiles of dialogue are useful tools for studying properties of defeasible reasoning and fallacies. It is explained how defeasible reasoning of the most common sort can deteriorate into fallacious argumentation in some instances. Conditions are formulated that can be used as normative tools to judge whether a given defeasible argument is fallacious or not. It is shown that three leading violations of proper dialogue standards for defeasible reasoning necessary to see how fallacies work are (1) improper failure to retract a commitment, (2) failure of openness to defeat, and (3) illicit reversal of burden of proof.

The straw man fallacy occurs where a critic misrepresents somebody's argument and then uses this misrepresented version (the so-called straw man) to refute the argument. Thus the straw man argument is a particular type of refutation of the kind studied in Chapter 2, except that the refutation is used in a fallacious way. Chapter 9 presents an analysis of the straw man fallacy defined as a misattribution of commitment in a Hamblin-style formal dialogue structure. The project undertaken in the chapter is to specify requirements for a commitment query device that can assist in making a fair ruling on straw man allegations. Three abstract models of commitment query inference engines are presented as objective methods of determining the content of an arguer's commitment store. This problem is vitally important to argumentation theory because it provides a test bed for helping us to extend and refine the notion of an arguer's commitment in dialogue.

Of all the problems studied in the book, except for the problem of applying schemes to real arguments, studied in Chapter 4, the problem posed by analyzing the straw man fallacy is the one where we are the farthest away from having a solution. As shown in the chapter, there already exist search engines in computer science that can partially solve the problem, but until software systems are extended to include defeasible argumentation schemes, there will be no adequate solution to the problem. It is argued nevertheless in Chapter 9 that the models we have so far are developed to a state of refinement where they can be used as guidelines to assist in dealing with problematic cases in which the straw man fallacy has allegedly been committed. There are currently implemented computational argumentation systems that use defeasible argumentation schemes, for example, Carneades, and so the problem set by the investigations of Chapter 8 is to apply these systems to the work done so far on the straw man fallacy. Hence the straw man fallacy investigated in Chapter 9 is an important avenue for future research on applying argumentation methods to significant problems in the field.

2

Argument Attack, Rebuttal, Refutation and Defeat

The aim of this chapter is to clarify a group of related terms, including 'argument attack', 'rebuttal', 'refutation', 'challenge', 'critical question', 'defeater', 'undercutting defeater', 'rebutting defeater', 'exception' and 'objection', which are commonly used in the literature on argumentation. The term 'rebuttal' is often associated with the work of Toulmin (1958), while the terms 'undercutting defeater' and 'rebutting defeater' are associated with the work of Pollock (1995) and are commonly used in the artificial intelligence literature. The notions of argument attack and argument defeat are associated with a formal model of argumentation that is prominent in artificial intelligence called the abstract argumentation framework. As shown in the chapter, these terms are, at their present state of usage, not precise or consistent enough for us to helpfully differentiate their meanings in framing useful advice on how to attack and refute arguments. An additional difficulty is that argument diagramming tools are of limited use if they cannot represent the critical questions matching an argumentation scheme. A way of overcoming both difficulties is presented in this chapter is by using the Carneades Argumentation System.

It is a widely accepted idea in recent models of argumentation in artificial intelligence that there are three ways of attacking an argument: (1) premise attack, (2) mounting another argument to attack the conclusion of the previous one that is the target of the attack, and (3) undercutting the previous argument, not by attacking its premises or its conclusion but by attacking the argument itself. Many of the critical questions matching argumentation schemes appear to represent this undercutter type of attack, which would make them fall into category 3, but in some instances asking a critical question, in order to refute the given argument, needs to be backed up by evidence. In these kinds of cases it seems natural to think that the asking of a critical question represents an attack that falls into category 2. However, as shown in Chapter 1, there are fundamental differences in the formal models of argumentation used in computing, as well in the argument visualization systems that they use, when it comes to representing these ways of attacking an argument. This poses some fundamental problems if the computational tools, including the argument diagramming tools developed in artificial intelligence, are to be used in conjunction with methods of analyzing and evaluating arguments by practitioners of informal logic. It is these problems that are taken up in Chapter 2.

Section 1 introduces some of the commonly accepted terminology. Section 2 presents a simple explanation of the basic idea of abstract argumentation frameworks purporting to formally model the notions of argument attack and defeat. Section 3 provides a brief explanation of the traditional philosophical notion of refutation exemplified by the Socratic-style refutation dialogue called the elenchus. Section 4 provides an extension of the example briefly introduced in Chapter 1 about whether video games lead to violence to illustrate the problems confronted in Chapter 2. This example introduces an important distinction between an internal refutation and an external refutation. This distinction, and the running example of its use, provides the departure point for the rest of the chapter. Sections 5–7 show how Carneades made it possible to represent critical questions matching an argumentation scheme by distinguishing three kinds of premises in a scheme, called ordinary premises, assumptions and exceptions. First, it is shown how Carneades did this using only the simpler type of argument diagram that does not have entanglement. Next is shown how Carneades moved to an improved solution to the problem by using entanglement to model exceptions by adopting the notion of a Pollock-style undercutter. Section 8 shows briefly how Carneades has the capacity for modeling another kind of objection to an argument, namely, the objection that the argument is irrelevant. Section 9 states the conclusions of the chapter. Section 10 provides a classification system to bring some order to the notions of argument attack, critical questioning, undercutting, rebuttal, internal refutation, external refutation and argument defeat.

1. Questions about Attack, Rebuttal, Objection and Refutation

One finds it to be a widely held commonplace in writings on logic and artificial intelligence that there are three ways to attack an argument (Prakken, 2010, 169). One is to argue that a premise is false or insufficiently supported. Let's call this the premise attack. Another is to argue that the conclusion doesn't follow from the set of premises that were presented as supporting it. This could be called an undercutting attack, as we will see below. The third is to argue that the conclusion is shown to be false by bringing forward a counterargument opposed to the original argument. What the attacker needs to do in such a case is to put forward a second argument that is stronger than the original argument and that provides evidence for rejecting the conclusion of the original argument. Such an attack is sufficient to defeat the original argument, unless its proponent can give further reasons to support it.

The undercutting type of attack does not apply to deductively valid arguments. If an argument fits the form of a deductively valid argument, it is impossible for the premises to be true and the conclusion false. Deductive reasoning is monotonic, meaning that a deductive argument always remains valid even if new premises are added. However, there is a method of attack on defeasible arguments that is highly familiar in the recent research on nonmonotonic logics for defeasible reasoning. It is to argue that there is an exception to the rule and that the given case falls under the category of this type of exception. This way of attacking an argument is very familiar in recent studies of defeasible reasoning, like the classic Tweety inference: birds fly; Tweety is a bird; therefore Tweety flies. This inference is based on the defeasible generalization that birds normally fly, or it could also be analyzed as being based on a conditional rule to the effect that if something is a bird it flies. Such a conditional is open to exceptions, meaning that it may default in some cases. The argument can be attacked by pointing out the exception to the rule.

To attack an argument in the third way, it may be enough to simply question whether its conclusion is true, but if a given argument that is being attacked has a certain degree of strength, merely questioning its conclusion may not be sufficient. What the attacker needs to do in such a case is to put forward a second argument that is stronger than the original argument and that provides evidence for rejecting the conclusion of the original argument. Such an attack is sufficient to defeat the original argument, unless its proponent can give further reasons to support it (Dung, 1995). Still another way to attack an argument is to ask a critical question that casts the argument into doubt and that may defeat the argument unless its proponent can make some suitable reply to the question. The form of attack will be taken up in Section 4.

Even though the given argument may stand, having repelled all attacks of the first three kinds, it may still be defeated on other grounds. One of these is that the argument is irrelevant, even though it may be valid. What is presupposed by this fourth kind of attack is that the given argument is supposed to be used to resolve some unsettled issue in a discussion that is being carried on in the given case. To attack an argument in the fourth way, matters of how the argument was used for some purpose in a context of dialogue need to be taken into account. If an argument has no probative value as evidence to prove or disprove the ultimate *probandum* in this particular discussion, it may be dismissed as irrelevant. Discussions of argument attack and refutation in the literature tend to acknowledge the first three ways of attacking an argument but to overlook the fourth way. The reason could be that this fourth way is more contextual than the first three ways in that it more directly relates to the context of dialogue surrounding the given argument. It could be classified as a procedural objection rather than as an attack.

Still another way to attack an argument is to claim that it commits the fallacy of begging the question. A circular argument, like 'Snow is white therefore snow is white', may be deductively valid but still be open to attack on the grounds that it fails to prove its conclusion. The failure here relates to the requirement that the premises of an argument that is being used to prove a conclusion should carry more weight than the conclusion itself. Thus if one of the premises depends on the conclusion, and cannot be proved independently of the conclusion, it is useless to increase the probative weight of the conclusion. Such an argument may be valid, but it is open to the criticism that it is useless to prove the conclusion it is supposed to be proving.

Although there may be four basic ways to attack an argument, asking a critical question is a way of making an objection to an argument that may or may not be seen as an attack on the argument. The notion of making an objection to an argument seems to be much broader than the notion of attacking an argument, for making an objection can be procedural in nature. We also need to be careful to note that there can be ways of making an objection to an argument that do not fall into any of these five categories of attack on an argument (Krabbe, 2007). Thus, the task of defining the notion of an objection precisely, and the task of classifying the various types of objections that can be made to an argument, remain open questions for future work. Still, in this section we have made some progress toward this investigation by carefully describing four basic ways to attack an argument and by adding that asking a critical question may also often be seen as a way of attacking an argument by raising critical doubts about it. Argument attacks surely represent some of the central ways of raising an objection about an argument.

Perhaps the best known use of the term 'rebuttal' in argumentation theory is Toulmin's use of it in his argument model, containing the elements datum, qualifier, claim, warrant, backing and rebuttal. In the model (Toulmin, 1958, 101), the datum is supported by a warrant that leads to a claim that is qualified by conditions of exception or rebuttal. For example (99), the claim that a man is a British subject might be supported by the datum that he was born in Bermuda, based on the warrant that a man born in Bermuda will be a British subject. The warrant appears to be similar to what is often called a generalization in logic. This example of an argument is defeasible, because the generalization is subject to exceptions, and hence the argument is subject to defeat if the information comes in showing that the particular case at issue is one where an exception holds. For example, although a man may have been born in Bermuda, he may have changed his nationality since birth (101). Toulmin uses the word 'rebuttal', but other words such as 'refutation' or 'defeater' might also be used to apply to such a case.

The meaning term of the term 'warrant' in Toulmin's argument layout has long been the subject of much controversy (Hitchcock and Verheij, 2006). A Toulmin warrant is in typical instances a general statement that acts as an inference license, in contrast to the datum and claim that tend to be specific statements. In logical terms, it could be described as a propositional function or open sentence of this form: if a person *x* was born in Bermuda, then generally that person *x* is a British subject.

A rebuttal, judging by Toulmin's Bermuda example, is an exception to a rule (warrant, in Toulmin's terms). However, according to Verheij (2009, 20), rebuttal is an ambiguous concept in Toulmin's treatment, and five meanings of the term need to be distinguished. First, rebuttals are associated with "circumstances in which the general authority of the warrant would have to be set aside" (Toulmin, 1958, 101). Second, rebuttals are "exceptional circumstances which might be capable of defeating or rebutting the warranted conclusion" (Toulmin, 1958, 101). Third, rebuttals are associated with the nonapplicability of a warrant (Toulmin, 1958, 102). But a warrant could also be an argument against the datum, a different sort of rebuttal from an argument against the warrant or the claim. In traditional logical terms, this would be an argument claiming that a premise of the inference being rebutted does not hold. Verheij also distinguishes between the warrant that acts as an evidential support of the conditional and the conditional that is one premise in the inference. On his analysis a rebuttal can attack the conditional or it can attack the warrant that supports the conditional as evidence.

Describing rebuttal as citing an exception to a rule of inference on which an argument was based sounds similar to what is called undercutting in the literature on defeasibility (Pollock, 1995). Pollock's distinction between two kinds of counterarguments called rebutting defeaters and undercutting defeaters (often referred to as rebutters vs. undercutters) is drawn as follows. A rebutting defeater gives a reason for denying a claim by arguing that the claim is a false previously held belief (Pollock, 1995, 40). An undercutting defeater attacks the inferential link between the claim and the reason supporting it by weakening or removing the reason that supported the claim. The way Pollock uses these terms, a rebutter gives a reason to show the conclusion is false, whereas an undercutter merely raises doubt as to whether the inference supporting the conclusion holds. It does not show that the conclusion is false. The classic example is the Tweety argument. If new information comes in telling us that Tweety is a penguin, the original Tweety argument is undercut. Generally speaking, the argument still holds. Generally birds fly, and, hence, given that Tweety is a bird, it follows that Tweety flies. But in this particular case, we have found out that Tweety is a penguin. Hence in this particular case, since we know that Tweety is type of bird that does not fly, we can no longer use the former inference to draw the conclusion that Tweety flies.

Pollock has another example (1995, 41) that illustrates a defeasible argument that could be called argument from perception.

For instance, suppose *x* looks red to me, but I know that *x* is illuminated by red lights and red lights can make objects look red when they are not. Knowing this defeats the prima facie reason, but it is not a reason for thinking that *x* is *not* red. After all, red objects look red in red light too. This is an *undercutting defeater.* (Emphasis in original)

To show how the red light example has the defining characteristics of a species of rebuttal, we can analyze it as an initial (given) argument and a counterargument posed against it. The original argument says: when an object looks red, then (normally, but subject to exceptions) it is red, and this object looks red to me, therefore this object is red. The rebuttal of the original acts as a counterargument that attacks the original argument: this object is illuminated by a red light, and when an object is illuminated by a red light, this can make it look red even though it is not, therefore the original argument (the *prima facie* reason for concluding that this object is red expressed by the original argument) no longer holds. According to Pollock (1995, 41) the counterargument should be classified as an under-cutter rather than a rebutter because red objects look red in red light too. Even given the attacking argument, the object may be red, for all we know. Thus in Pollock's terms it would not be right to say that the attacking argu-ment is a rebutting defeater that shows that the conclusion of the original argument is false. What it shows is that because of the new information about the red light, the counterargument, built on this new information, casts doubt on the conclusion of the original argument. As an undercutter it acts like a critical question that casts an argument into doubt.

Pollock's distinction between rebutters and undercutters is clearly fun-damental to any understanding of defeasible reasoning, but from a practi-cal point of view, it leaves a number of questions open. Is an undercutter a particular instance that makes a defeasible generalization fail in a specific case? Or is an undercutter a special type of counterargument that attacks a prior defeasible argument and acts as a rebuttal to it? Is there a special characteristic of the logical structure of defeasible arguments that leaves them open to an undercutter type of attack, and if so how can we identify this characteristic so that we can learn when it is appropriate to make an undercutter type of attack? These are all practical questions that might be helpful in telling a participant in argumentation, or a critic of an argument, how to attack that argument or critically question it by finding some sort of standard rebuttal that applies to it.

There are also some terminological questions about how to classify the terms 'attack', 'rebuttal' and 'refutation'. Pollock's terminology can be somewhat confusing when we try to apply it to giving practical advice on how to attack, rebut, critically question or refute a given argument, because undercutting does not sound all that different from rebutting. If I find an exception to a rule that defeats the defeasible argument, as in the red light example, surely it is reasonable to say that I have attacked or even rebutted the original argument. How is rebuttal different from refutation, a term often used in logic textbooks and writings on logic over the centuries? Currently, the terms 'attack' and 'defeat' are being widely used in writings on argumentation in artificial intelligence and on how these terms fit into the picture.

2. Abstract Argumentation Frameworks

There is a formal model of argumentation currently being widely applied in artificial intelligence that is built around the idea of analyzing and evaluating argumentation on the basis of how one argument attacks another. This influential way of formally modeling argumentation is called an abstract argumentation framework (Dung, 1995). It seems like a natural model, because argumentation by its very nature evaluates arguments by looking at both sides of an issue and weighs the pro arguments against the contra arguments, the stronger arguments defeating the weaker ones. Looking at argumentation in this way, the process is one of judging argument to be strong or weak on the basis of how strong the counterarguments are that go against it, to see whether the original argument can stand up to these counterarguments or not. It does not try to define the notion of one argument attacking another, but again takes this relation as primitive. An abstract argumentation framework (*AF*) is defined as a pair $\langle Args, Def \rangle$, where *Args* is a set of arguments and $Def \subseteq Args \times Args$ is a binary relation of defeat. The model does not reveal anything about the internal construction or parts of an argument (its premises and conclusion, or the nature of the inferential link from the premises to the conclusion). The other primitive notion is that of argument defeat.

Argumentation is evaluated by forming a sequence in which a second argument attacks the first one, and then a third argument attacks the second one, and so forth. This process is repeated until it runs out of arguments. An argument is acceptable if it is not defeated by any other argument. An argument is not acceptable as soon as it is defeated by any other argument. Frequently, the language of 'in' and 'out' is used to describe this process of argument evaluation. An argument is said to be *in* if all its defeaters are out. An argument is said to be *out* if it has even one defeater that is in. The easiest analogy to understand how this process works is that of a close-range gunfight like the legendary Gunfight at the O.K. Corral. A participant is still in if he is not shot by any of the other participants. A participant is out if he is shot by even one of the other participants. In other words, the assumption is that every shot is fatal, or at least deadly enough to knock the participant out of the gunfight.

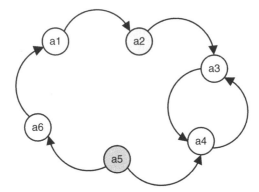

FIGURE 2.1 First Step of the Argumentation Sequence

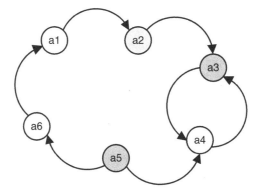

FIGURE 2.2 Second Step of the Argumentation Sequence

Let's start with a simple example of a Dung-style argument diagram to illustrate how a sequence of argumentation would be evaluated. As shown in Figure 2.1, there is an argument structure containing six arguments labeled a1 through a6. A node with no shading inside it is neither in nor out. A node with light shading represents an argument that is in. A node containing darker shading represents an argument that is out.

In the example shown in Figure 2.1, argument a2 is in, while argument a5 is out. So what happens next? Since a3 is attacked by an argument that is in, namely, a2, a3 is out. This is shown in Figure 2.2, where a3 is contained in a darkened node.

But now, since a3 is out, and since we already knew that a5 is out, we know that a4 is not attacked by any argument that is in. Therefore a4 has to be in, so a4 is shown in a node with lighter shading in Figure 2.3.

Now what happens? Since a6 is attacked only by a5, and since a5 is out, a6 is not attacked by any argument that is in. Therefore a6 is in. This result is shown in Figure 2.4.

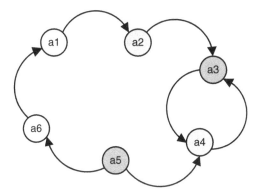

FIGURE 2.3 Third Step of the Argumentation Sequence

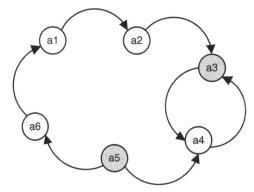

FIGURE 2.4 Fourth Step of the Argumentation Sequence

Now that we know this much about the other arguments, we can reach a final conclusion about argument a1. Since a1 is attacked only by a6, and since a6 is in, a1 is out. This final outcome once all the arguments have been taken into account is shown in Figure 2.5.

There is more to abstract argumentation frameworks, but this simple example can illustrate basically how the system works.

The question to be raised is whether abstract argumentation frameworks can provide a method for analyzing and evaluating argumentation. There seem to be many who think it can, but there are also reasons to think that it may be limited in its capability to provide such a method. Many commentators would be likely to reply that surely evaluating an argument cannot be exclusively carried out by just checking to see whether all the known counterarguments to it have been refuted. Surely, an argument evaluator also needs to see whether there is some positive support for the argument. For according to the central methodology of argumentation one needs to weigh the counterarguments against the pro arguments that offer positive

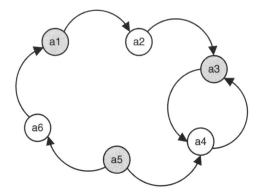

FIGURE 2.5 Fifth Step of the Argumentation Sequence

support for a claim to see if the claim is adequately supported by weighing the pro against the contra. In other words, the formal abstract argumentation model, if it is expected to provide the exclusive basis for a method of analysis and evaluation of argumentation, seems to emphasize the negative aspect too much by concentrating exclusively on argument attacks as opposed to also taking into account the positive evidence that supports a particular argument. To propose that an argument can be judged to be justified and therefore accepted if and only if it survives conflict with all the known counterarguments seems too weak and negative as a general procedure for evaluating arguments. Just because an argument survives conflict with all counterarguments would not seem to imply that it should be evaluated as a good or strong argument, meaning that it is an argument that is acceptable. It must also have positive support from other arguments.

In general, it seems reasonable to hold that any analysis of argumentation, or formal model of argumentation, must take into account both the pro and contra arguments with respect to a given thesis that is at issue. In the abstract argumentation model, the nodes represent whole arguments, and there needs to be a transition from this model to a more complex model with an additional structure to combine pro and contra arguments. This larger formal model of argumentation structure needs to take positive supporting arguments into account as well as the attacking arguments directed against the claim at issue.

As a formal framework of argumentation, the abstract argumentation formalism does, however, have a discernible dialogue structure that represents argumentation of a particular kind. In this type of dialogue there are two participants, a proponent and an opponent, who take turns engaging in argumentation with each other. The proponent starts the procedure by putting forward a particular argument. The opponent moves next by attacking the proponent's argument with the counterargument. The proponent moves next by attacking the opponent's argument using another counterargument. And the dialogue proceeds in precisely this fashion by move and countermove. It is a restrictive form of dialogue in that no other moves, for example, speech acts in the form of asking a question, are allowed, and no deviations from the turn-taking procedure of attack and counterattack are permitted. The procedure continues until one party or the other cannot make a further move. The sequence of argumentation is evaluated with the decision made by an external referee that some of the arguments in the sequence are accepted (in), while other ones are not accepted (out). Once this determination has been made by the referee, the sequence of argumentation can be made using the abstract argumentation model, as illustrated by the example. By applying this procedure it can be determined whether the very first argument put forward by the proponent (at move 1) is in or out. If it is in, the proponent has won. If it is out, or if it is neither in nor out, the respondent has won.

3. Socratic Refutation Dialogues

There are known instances of this type of dialogue that have been studied, and indeed the study of argumentation fitting this format has a long history stemming from studies on argumentation in ancient Greek philosophy and logic. One example can be given here. As a simple example to illustrate how a formal dialectical structure in the form of the game can be used to illustrate argumentation in a dialogue setting, Hamblin (1971) chose a medieval dialectical game. Argumentative dialogue games of this sort were studied in the Middle Ages, deriving from the Socratic dialogues of Plato and the writings on dialectic by Aristotle. The example chosen and formalized by Hamblin (1971, 260–264) is called the Obligation game. The dialectical structure of the procedure is formulated by Hamblin in a technical manner, but the essence of it can be described quite simply in outline as a ' game. The game is played by two participants, called the opponent and the respondent. The opponent speaks first and puts forward a particular proposition that is posited. The respondent moves next and has only two choices of moves. He can repeat the preceding location of the opponent, or he can state its negation. In other words, he can agree with it or disagree with it. The respondent has a commitment store consisting of the original proposition that was posited, and all his answers to date. The opponent wins if he can find an inconsistency in the commitment set of the respondent. The respondent wins if he has survived some given number of moves that is stipulated in advance without committing himself to an inconsistency. This game illustrates a typical Hamblin structure of dialogue, in that there are two parties who take turns making moves in the form of locutions, or speech acts as we would now call them, and propositions are inserted into the commitment sets of the parties as each move is made, according to the commitment rules of the game. Hamblin's treatment of the game also illustrates a typical feature of writings on logic at the time, namely, its use of deductive logic to model the inferential structure of how propositions are derived from other propositions in the game.

It is clear that the Obligation game was meant to model, at least in a simple format, the pattern of argumentation called the elenchos or elenchus, a sequence of moves of the kind often found in the Platonic dialogues in which Socrates examines an opinion put forward by another party. Socrates' technique is to ask his respondent to answer a sequence of questions requiring the respondent to make a choice to commit himself to a series of propositions, and to show by logical reasoning that the set of propositions the respondent has committed himself to contains an inconsistency or some other sort of absurdity. The aim of the procedure is not to make the respondent look ridiculous, but to show that his original opinion is not tenable. A deeper aim of the procedure is to explore some controversial issue by probing into the arguments for and against a controversial proposition in order to deepen our philosophical understanding of the issue.

According to Robinson (1953, 7), the term *elenchus* has both a wider and narrower meaning. In the wider sense, it means "examining a person with regard to a statement he has made, by putting to him questions calling for further statements, in the hope that they will determine the meaning and truth-value of his first statement". However, since falsehood is the truth value most often expected, the elenchus has a narrower meaning, referring to a kind of refutation. It can take the form of a cross-examination in a legal setting, but in a trial the argumentation is directed toward a judge or jury. In a Socratic dialogue the elenchus is directed specially to one person whose individual opinion is being examined. It is this latter special sense that fits the technique so often used by Socrates in the Platonic dialogues, where he traps the respondent into a contradiction by asking a series of questions that leads him into admitting something that is the opposite of his original opinion on some controversial issue.

On Robinson's account of it, this Socratic sequence of questioning and answering proceeds through five phases.

1. First, Socrates asks his respondent some general question, very often in the field of ethics. This question is a source of doubt and difficulty, as it represents a controversy on which there are strong opinions on both sides.
2. Second, the respondent answers the question, by taking one side on the issue in a manner that clearly indicates that he is advocating his opinion of something in which he believes and sincerely accepts as true.
3. The third phase is that Socrates puts a series of secondary questions to the respondent. These secondary questions are different from the primary question in that the answers seem obvious to the respondent, and usually the answer that is called for is yes. These secondary

questions are typically what might be called leading questions, in that offering anything other than the intended answer seems odd and might be hard to defend. As Robinson (1953, 7) describes them, these questions are not requests for information but can be better described as "demands for an assent that cannot very well be withheld". In a typical instance of the Socratic elenchus, these secondary questions fall into groups, and it is unclear how they relate to the primary question. During this sequence of questioning it tends to be unclear to the respondent and to the readers of the dialogue where the line of argumentation is going.

4. The fourth phase is the end point of the sequence of questioning where Socrates puts all the respondent's concessions together and shows that it leads by a chain of logical reasoning to his answer to the primary question. At the end of this fourth phase, then, it has become apparent that the respondent has contradicted himself. Robinson puts it this way: "Propositions to which the answerer feels he must agree have entailed the falsehood of his original assertion".

5. The fifth phase is the aftermath. The respondent is bewildered by this unexpected outcome, and it makes him feel upset, even ashamed, that he has done something so embarrassing as to contradict himself on an issue about which he felt so strongly convinced of the truth of his opinion at the beginning (*Meno*, 80ab).

This outcome seems negative, but Socrates justifies it by his theory that the only way to attain knowledge about something is to first of all start with the awareness that you are ignorant about it. The distinctive aspect of Socrates' philosophy is that he does not claim to have knowledge, and only claims to be wiser than other people who have not reflected about philosophical issues on the ground that he at least knows that he does not know. In an interesting way, this philosophy reflects earlier skeptical views of Greek philosophers who claimed that we as humans do not have knowledge of the truth that is not potentially subject to error or distortion. In order to achieve this end of getting to knowledge by convincing of ignorance, the elenchus has to be a "very personal affair" (Robinson, 1953, 15), in which the respondent is convinced of the logical validity of the chain of reasoning once the procedure has reached its end point.

Hamblin's formalization of the Obligation game is fundamentally interesting, from the point of view of the formalization of argumentation structures, because it illustrates his basic idea that the best way to study arguments is to frame them in a dialogue setting in which several parties – in the simplest instance, two – take turns making moves governed by rules that determine which propositions each party may rightly be said to be committed to in virtue of these rules and moves previously made. Caminada (2008) has provided a formalization of Socratic elenchus using

TABLE 2.1 *Sequence of Dialogue Moves in an Elenchus*

Sequence from Start to End	Socrates	Respondent
Primary Question	Asks question on some issue	Gives answer *A* on his opinion
Secondary Questions Begin	Asks first leading question	Gives suggested answer
	Asks second leading question	Gives suggested answer
	Asks last question in this group	Gives suggested answer
Second Group of Questions	Asks first question in next group	Gives suggested answer
	Asks more questions	Gives suggested answer
Last Group of Questions	First question of last group	Gives suggested answer
	Last question of last group	Gives suggested answer
Endpoint of Questioning	Concessions all put together	Expresses agreement
Contradiction Revealed	Primary answer leads to not-*A*	Respondent shocked
Aftermath	Ignorance revealed	Respondent now wiser

an abstract argumentation formal structure to represent the sequence of moves in the dialogue. In this structure, the chain of argumentation starts with a proposition put forward by one party in a dialogue and proceeds by an absurdity being derived from it by the other party. A main difference between this structure and the Hamblin structure of the Obligation game is that in Caminada's formalization, the propositions from the commitments of the first party are defeasible consequences of them.

Some questions are raised, however, on how well either formalism represents the sequence of moves in a Socratic-style refutation dialogue. We can see from Robinson's analysis of the five stages of this questioning sequence that any realistic example of Socratic dialogue of this sort groups the moves together in a way different from either formal model. Table 2.1 gives us a general outline of how this sequence of questioning and answering generally goes, on Robinson's description of it, from the asking of the primary question at the beginning to the end point where the contradiction is revealed, and finally to the aftermath.

Once the primary question has been asked, then we go into the sequence of secondary questions. These fall into groups, and can be ordered into a first group, second group, and so forth. There can be any number of groups of secondary questions. Once all of them have been asked we reach the end point of questioning where Socrates puts all the concessions together. At this stage Socrates has to explain to the respondent how all these concessions fit together and how they lead by logical reasoning to a particular proposition that is the opposite of the respondent's primary concession. Once the respondent grasps the contradiction, the end point of the elenchus as a sequence of logical reasoning from the beginning point to the end point has been reached. The aftermath is only an add-on that indicates the educational effect of the sequence that Socrates postulates according to his theory of learning.

So here we have some tantalizing suggestions. The Socratic type of argumentation does indeed represent some notion of refutation. But is it the general notion of refutation that we want to use for purposes of argumentation studies generally? The Obligation game notion of refutation modeled by Hamblin's formal system does represent, to some extent at least, the logical structure of a notion of refutation where one party refutes another by deriving logical implications from the commitments agreed to by the first party. But we can also ask how well this notion of refutation generalizes to the notion of refutation we need for argumentation studies. Some of the same remarks about applicability can be raised in regard to the use of abstract argumentation models to represent the notion of argument defeat. In the literature on abstract argumentation, the notions of 'attack' and 'defeat' are taken to represent the same primitive concept of defeat, one argument attacking another in an abstract argumentation structure. So while we seem a little further ahead of the quest to understand the notions of argument rebuttal and refutation, we are still by no means in a good position to draw clear distinctions among the family of concepts represented by the terms 'argument attack', 'rebuttal', 'refutation' and 'defeat'.

One idea, however, that has been fundamental to the notion of refutation both in Hamblin's formal dialectical model of the Obligation game and in the abstract argumentation of Socratic refutation provided by Caminada is that an arguer's argument is refuted by showing that it is inconsistent with his set of prior commitments in a dialogue. However, this type of refutation is different from showing that the arguer's argument is defeated by external evidence that shows that the argument is no longer tenable. These remarks suggest that a distinction can be drawn between internal and external refutation of an argument.

4. Internal and External Refutation

Goodwin (2010) presented a methodical procedure to her students on how to refute an argument that contrasts two strategies. The first strategy is that of focusing on the argument's conclusion and arguing for the opposite. She offered the following example. If one side argues that video games lead to violence, the other side can argue that video games do not lead to violence. This can be recognized as a strategy often called rebuttal or refutation. It is the strategy when confronted with a target argument to present a new argument that has the opposite (negation) of the target argument as its conclusion. Although conceding that this is an important and often effective strategy, she suggests another one that may be even better. Instead of just looking at the conclusion of the other argument, this second strategy is to examine the reasons the other side is giving to support its argument and to see if these reasons hold up under questioning. Among the questions she proposed as ways of attacking the other argument are (1) to ask whether the other side is relying on a biased source, (2) to ask whether the evidence the other side is citing is relevant, or (3) to ask whether the analogy put forward by the other side is really similar.

What is suggested by this advice is that there are basically two ways of attacking an argument. One way, generally called refutation, is to present a new argument that has as its conclusion the negation of the original argument. Below we will challenge this generally accepted meaning of the term 'refutation' on the grounds that it is too broad. The problem is that we often have cases where a new argument has as its conclusion the negation of an original argument, but the new argument might still be weaker than the original argument. In such cases it is questionable whether the new argument is a refutation of the original one. For the moment, however, we accept the broad conventional meaning of the term 'refutation' as a point of departure. The other way of attacking an argument, generally called asking critical questions or casting doubt on an argument, is to ask questions that relate to the particular form of the original argument. For example, if the original argument was based on a source, such as witness testimony or expert testimony, one could ask the critical question of whether that source is biased. Or if the original argument has the form of an argument from analogy, one could ask the critical question of whether the two cases at issue are really similar. Goodwin states that although attacking the other side's reasons by asking critical questions involves more strategy and paying attention to what the other side says, it can often be more effective because it attacks the opposed argument internally, nicely causing it to fall down.

This practical advice on how to refute an argument is generally very interesting from the point of view of argumentation theory, because it suggests there are two distinctive strategies – refutation and critical questioning, as each might be called – that need to be separated and that each calls for a different approach. She has shown that each type of argument strategy has a distinctively different structure from the other. This is an important distinction for argumentation theory. Hamblin (1970, 162) distinguished between a weaker and a stronger sense of the term 'refutation'. The weaker he describes as "destruction of an opponent's proof" and the stronger as "construction of the proof of a contrary thesis". It would be nice to have some terminology to make this important distinction between these two meanings of the term 'refutation'. Let us call destruction of an opponent's proof internal refutation, because, as Goodwin has described it, this strategy is to examine the reasons the other side is giving to support its argument and to see if these reasons hold up under questioning. It is an internal attack on the argumentation offered by the other side. Let us call the construction of the proof of a contrary thesis external refutation because it goes outside the original argument to present a new argument that has as its conclusion the negation of the original argument. Attacks can be internal or external.

An example she gives to illustrate the technique of internal refutation is quoted below (with some parts deleted). This was the example on which Figure 1.2 was based. The other side takes the view that video games do not lead to violence.

The other side said that Dr. Smith's study clearly shows that video games do not lead to violence. But Dr. Smith is biased. His research is entirely funded by the video game industry. That's what the 2001 investigation by the Parent's Defense League demonstrates. So you can see that the other side has no credible evidence linking video games to violence.

In the example one can see the components of a refutation. First, there are two parties that are presenting arguments on opposed sides of the disputed issue. The issue is whether or not video games lead to violence. The first side has argued that video games do not lead to violence, and has supported its claim by bringing forward the evidence that Dr. Smith's study shows that this claim is true. The opposed side then presents a counterargument, but this counterargument is not an external refutation, a new argument that supports the claim that video games do lead to violence. Instead, it attacks the original argument internally by making the claim that Dr. Smith is biased, and supports it with the reason that his research is entirely funded by the video game industry. So this is a counterargument, but not a refutation in the sense defined above. It is something else. It corresponds to the other technique of attacking an argument that Goodwin described as attacking the reasons the other side is giving by asking critical questions.

We can even analyze this internal type of attacking strategy more deeply by pointing out that the original argument took a particular form. It appears to be an argument from expert opinion that cites a study by someone called Dr. Smith that supposedly showed that video games do not lead to violence. The field of expertise of Dr. Smith is not stated, but it appears we are meant to assume that Dr. Smith is an expert in some field that includes the study of whether video games lead to violence or not. If we can make this assumption, the form of the original argument can then be identified as that of argument from expert opinion. Given this assumption we can understand a little more about the structure of the internal attack used against this argument. The attack makes the claim that Dr. Smith is biased, and this particular type of attack undercuts the argument by finding a weak point in its structure that, once pointed out and supported by evidence, subjects the argument to doubt in such a way that it no longer holds up as a way of supporting its conclusion that video games do not lead to violence.

Now a general problem is posed. Should asking a critical question be viewed as an argument that undercuts the second? In other words is asking this critical question a rebuttal of the original argument from expert opinion, or merely an objection that raises doubt about whether the argument from expert opinion originally put forward still holds? Questions are different from statements, and while statements can be represented in text boxes as premises and conclusions on the standard argument diagram, there is no straightforward way to represent a question in this manner on an argument diagram. One way to do this will be now be investigated, where the statement that Dr. Smith is biased is represented as an undercutter, a counterargument that attacks the original argument by questioning whether the inferential link from the premises to the conclusion genuinely holds in the instance in question even though the argument fits a genuine argumentation scheme. One of the problems in the literature on argumentation in artificial intelligence is how to model this type of argument by representing its structure on an argument map.

5. Argumentation Schemes and Critical Questions

Pollock's red light example can be fitted to an argumentation scheme that has been called argument from appearance (Walton, 2006a). Although Pollock did not employ the concept of an argumentation scheme with matching critical questions, the pattern of inference of the red light example can be called argument from perception (Walton, Reed and Macagno, 2008, 345).

Premise 1: Person P has a φ image (an image of a perceptible property).

Premise 2: To have a φ image (an image of a perceptible property) is a *prima facie* reason to believe that the circumstances exemplify φ.

Conclusion: It is reasonable to believe that φ is the case.

Walton, Reed and Macagno (2008, 345) list this form of argument as an argumentation scheme with the following critical question matching it: are the circumstances such that having a φ image is not a reliable indicator of φ?

Consider another example (Prakken, 2003): if something looks like an affidavit, it is an affidavit: this object looks like an affidavit, therefore it is an affidavit. This inference might fail if we are taking part in a television series about a trial in which props are used. A document on a desk might look like an affidavit, but after all, this is a TV series. It might not be an affidavit, but merely a prop made to look like one. In the context, the original argument fails to support the conclusion that the document in question is an affidavit. But maybe it is a real affidavit. An easy way to get such a prop for the TV series would be to ask someone who has access to real affidavits to get one for use in the TV series. This example has the same scheme as the red light example.

The scheme representing argument from expert opinion was described in Chapter 1, along with its matching set of critical questions. An argument from expert opinion needs to be evaluated in a dialogue where an opponent (respondent) can ask critical questions. This form of inference is defeasible, provided we take it to be based on a defeasible generalization to the effect that if an expert says A, and A is in the right field for the expert, then A may plausibly be taken to be acceptable as true (subject to exceptions). What kinds of exceptions need to be taken into account corresponding to critical questions matching a scheme? Here we review the six basic critical questions matching this scheme.

CQ$_1$: *Expertise Question:* How knowledgeable is E as an expert source?

CQ$_2$: *Field Question:* Is E an expert in the field F that A is in?

CQ$_3$: *Opinion Question:* What did E assert that implies A?

CQ$_4$: *Trustworthiness Question:* Is E personally reliable as a source?

CQ$_5$: *Consistency Question:* Is A consistent with what other experts assert?

CQ$_6$: *Evidence Question:* Is E's assertion based on evidence?

CQ$_1$ refers to the expert's level of mastery of the field F. CQ$_4$ refers to the expert's trustworthiness. For example, if the expert has a history of lying or is known to have something to lose or gain by saying A is true or false, these factors would suggest that the expert may not be personally reliable. The assumption made in Walton (1997) is that if the respondent asks one of the six critical questions, the initiative shifts back to the proponent's side to respond to the question appropriately. The asking of the critical question defeats the argument temporarily until the critical question has been answered successfully. This approach is a first pass to solving the problem of how to evaluate an argument from expert opinion. More specifically, it is designed to offer students in courses on critical argumentation some direction on how to react when confronted with an argument from expert opinion.

The study of attacks, rebuttals and refutations would be aided considerably if some structure could be brought to bear that would enable us to anticipate in a particular case what sort of attack an argument is susceptible to. Here the critical questions matching a scheme can be very useful. For example, if the argument is an appeal to expert opinion, we can see already from examining the critical questions matching scheme for argument from expert opinion that this argument will tend to be open to certain types of attack. For example, it will be open to an attack on the grounds that the expert is not a trustworthy source. One of the standard ways of arguing that an expert is not a trustworthy source is to allege that the expert is biased because he or she has something financially to gain by making the claim. However, it has been shown that critical questions differ in their force. In some instances, merely asking a critical question makes the original argument default (be defeated), while in other instances, asking the critical question does not make the argument default unless the question asker can offer evidence to back up the question (Walton and Godden, 2005). There are differences between the critical questions on how strongly or weakly asking the question produces such a shift of initiative. Such observations have led to two theories about requirements for initiative shifting when critical questions matching the argument from expert opinion are asked (Walton and Godden, 2005). According to one theory, in a case where the respondent asks any one of these critical questions, the burden of proof automatically shifts back to the proponent's side to provide an answer, and if he or she fails to do so, the argument defaults. On this theory, only if the proponent provides an appropriate answer is the plausibility of the original argument from expert opinion restored. According to the other theory, asking a critical question should not be enough to make the original argument default. The question, if questioned, needs to be backed up with some evidence before it can shift any burden that would defeat the argument.

6. Managing Critical Questions with Carneades

Part of the definition of a rebuttal is that it is an attack on an argument, and a rebuttal itself would normally seem to be an argument. In order to define the notion of a rebuttal, we also need to have some clear notion of what an argument is. There is not much agreement in argumentation theory on how to define an argument, however. To cope with this problem, it is best to begin with a minimalist account of the structure of an argument. According to this account, an argument is composed of three things: a set of premises, a conclusion and an inference that leads from the premises to the conclusion. The conclusion is generally taken to be a claim that has been made, and the premises are propositions that are put forward in support of the claim. Beyond this minimal account, it will prove to be useful to have a formal model to represent the notion of an argument, preferably one that would enable us to visualize the premises and conclusion of an argument in a clear way to represent examples of attacks, rebuttals and refutations. For example, if we could represent Goodwin's example of an internal refutation, this capability could be extremely helpful. There are many such argumentation visualization tools available at the present time, but it is especially helpful to use one that provides not only a formal model of argumentation, but also an argument visualization tool that fits the model.

Argumentation is modeled by Carneades in a tree structure where the nodes are text boxes containing premises and conclusions of an argument (Gordon, 2010). The premises are connected to the conclusion in the normal way in an argument with an arrow pointing to the conclusion. An argument that supports a conclusion is indicated by a circle containing a plus

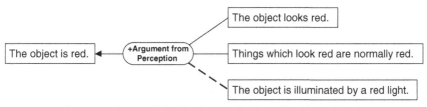

FIGURE 2.6. Exception Modeled by Carneades in Pollock's Red Light Example

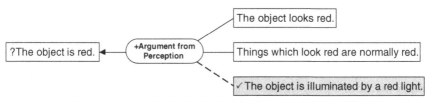

FIGURE 2.7 Undercutter in Pollock's Red Light Example Modeled by Carneades

sign. The premise is an exception, shown as joined to a circle by a dashed line. How Carneades displays the structure of the argument in Pollock's red light example is shown in Figure 2.6.

As shown in Figure 2.6, the statement at the bottom right is an exception, and so the argument as a whole represents a Pollock-style undercutter. In the Carneades model, this argument is represented as a typical defeasible argument that has two normal premises, displayed as the top two boxes on the right in Figure 2.6. But this argument is subject to an exception, and in Carneades the exception is represented as an additional premise of a special kind that can defeat the original argument. Carneades can also be used for evaluating arguments, and how the procedure of evaluation can be illustrated using the case of Pollock's red light example is shown in Figure 2.7. As shown by the checkmark text box at the bottom, the statement that the object is illuminated by red light has been accepted. Once the statement has been accepted, even though the two premises above it would normally enable the conclusion to be accepted provided these two premises are accepted, in this situation, since the exception applies, the conclusion is cast into doubt. The status of the conclusion is represented by the question mark appearing in its text box. This analysis visualizes a situation in which the conclusion is rendered questionable, and hence not acceptable. It does not tell us, however, that the conclusion is false or unacceptable.

Carneades defines formal properties that are used to identify, analyze, construct, visualize and evaluate arguments (Gordon and Walton, 2006a). Part of the definition of a rebuttal is that it is an attack on an argument, and a rebuttal itself is also an argument. It follows that in order to define the notion of a rebuttal, we surely also need to have some clear notion of what an argument is. As noted just above in this section, an argument is taken to have three basic components: a set of premises, a conclusion and an inference that leads from the premises to the conclusion.

Figures 2.6 and 2.7 show how these three components are related. In the following formal definition of an argument in Carneades (Gordon and Walton, 2009), a distinction is drawn between two types of opposition. One is negation, represented in the same way as in classical propositional logic where a proposition p is true if and only if its negation is false. The negation of a proposition, in other words, has the opposite truth value of the original proposition. The other is complement. The complement of a set is the set of things outside that set (Gordon and Walton, 2009, 242–243).

Definition: Let L be a propositional language. An *argument* is a tuple $\langle P, E, c \rangle$ where $P \subset L$ are its *premises*, $E \subset L$ are its *exceptions* and $c \in L$ is its *conclusion*. For simplicity, c and all members of P and E must be literals, that is, either an atomic proposition or a negated atomic proposition. Let p be a literal. If p is c, then the argument is an argument *pro p*. If p is the complement of c, the argument is an argument *con p*.

According to this definition we can understand the notions of an argument pro a proposition p and argument con a proposition p as follows. If p is the conclusion of the argument, the argument is said to be *pro p*, whereas if some proposition other than p is the conclusion of the argument, the argument is said to be *con p*. Defeaters (rebuttals) are modeled as arguments in the opposite direction for the same conclusion. If one argument is pro the conclusion, its rebuttal would be another argument con the same conclusion. Premise defeat is modeled by an argument con an ordinary premise or an assumption, or pro an exception (Gordon, 2005, 56). In the Carneades system, critical questions matching an argument are classified into three categories: ordinary premises, assumptions or exceptions. External refutations are modeled as arguments in the opposite direction for the same conclusion. If one argument is pro the conclusion, its refutation would be another argument con the same conclusion. Premise defeat is modeled by an argument con an ordinary premise or an assumption, or pro an exception (Gordon, 2005, 56). Seeing how Carneades models the distinction between internal and external refutation, we show how this distinction works in the case of argument from expert opinion.

Let's begin with the notion of external refutation to see how it works generally in cases of argument from expert opinion. In a case of external refutation, as shown in Figure 2.8, we have one argument from expert opinion in which the premise is that expert 1 says that some proposition A is true and the conclusion is the proposition that A is true.

This case is a special instance of the argument about Dr. Smith and Dr. Jones shown in Figure 1.7. The argument shown at the top in Figure 2.8 is a pro argument, as shown by the + in the circle representing the argument. Not quite an argument, for beneath it is the second argument that attacks the first argument, based on the premise that there is another expert who

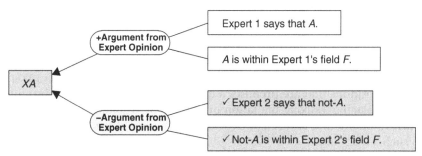

FIGURE 2.8 How Carneades Models External Refutation

says that the opposite of *A* is true. The second argument is an external refutation of the first one, because it is a separate opposed argument that has the opposite conclusion of the first argument.

But if the second argument is merely a rebuttal of the first argument can it properly be called a refutation? Certainly it fits the definition of an external refutation of the kind attributed to Hamblin above, but there is more to say about it. Notice that in Figure 2.8, the premises shown at the bottom appear in darkened textboxes and have check marks in front of them, indicating that this premise has been accepted. Notice that the premises shown at the top appear in undarkened text boxes with no check mark in front, indicating that each premise has merely been stated but has not been accepted. What will happen automatically in Carneades is that the bottom argument will be taken as refuting the top one. Since it has two accepted premises, when both premises are considered together, the conclusion *A* comes out as rejected (indicated by *X*).

In such a case, we can say that the first argument is refuted by the second one in a strong sense of the term 'refutation' meaning not only that the second argument goes to the opposite conclusion of the first one, but it does so in such a way that it overwhelms the first argument, providing a reason to infer that the conclusion of the first argument is no longer acceptable. We could say that in this strong sense of refutation, the second argument successfully refutes the first argument. Or perhaps we could draw the distinction in a different way by saying that the second argument not only rebuts the first argument but also refutes it. The terminology remains uncertain here but we will clarify it later.

No matter how we describe what has happened in this example in terms of the distinction between rebuttal and refutation, we can see why it illustrates how Carneades models the notion of an external refutation. In an external refutation, we have two separate arguments, and one attacks the other externally by providing an independent line of argument that goes to the opposite of the conclusion of the first argument. Carneades models the notion of an internal refutation in a completely different way by focusing on the critical questions matching the argumentation scheme, and goes into considerations of different ways these critical questions can be used to attack the original argument.

One of the main features of Carneades is that it enables critical questions to be represented on argument diagrams (Walton and Gordon, 2005). In the standard argument diagrams, the text boxes (nodes in the tree) contain propositions that are premises and conclusions of arguments, but the problem placing a strong limitation on the use of argument diagramming as an argumentation tool is that critical questions could not be represented on such a diagram.

Carneades solved this problem by enabling a distinction to be drawn between two ways an argument from expert opinion should be critically questioned, and thus enables the critical questions to be represented as implicit premises of an argumentation scheme on an argument diagram. The two assumptions – that (1) the expert is not trustworthy and that (2) what he or she says is not consistent with what other experts say – are assumed to be false. It is assumed, in other words, that (1) and (2) are false until new evidence comes in to show that they are true. The two assumptions – that (1) the expert is credible as an expert and that (2) what he or she says is based on evidence – are assumed to be true, until such time as new evidence comes in showing they are false. Also assumed as true are the ordinary premises that (1) the expert really is an expert, (2) he or she is an expert in the subject domain of the claim, (3) he or she asserts the claim in question, and (4) the claim is in the subject domain in which he or she is an expert.

Now let's look once again at the expertise question, to see how it could be classified. It is about E's depth of knowledge in the field F that the proposition at issue lies in. As noted above, the expertise question seems to ask for a comparative rating. What if the proponent fails to answer by specifying some degree of expertise, like "very credible" or "only slightly credible"? As noted above it seems hard to decide what the effect on the original argument should be. Should it be defeated or merely undercut? It seems like it should only be undercut, because even if we don't know how strong the argument from expert opinion is, it might still have some strength. It might even be very strong, for all we know.

The field and opinion questions can be modeled as ordinary premises of the arguments from expert opinion scheme in Carneades. Now let's look back at the trustworthiness question, which refers to the reliability of the expert as a source who can be trusted. If the expert was shown to be biased or a liar, that would presumably be a defeater. It would be an *ad hominem* argument used to attack the original argument and, if strong, would defeat it. But unless there is some evidence of ethical misconduct, as noted above, the proponent could simply answer 'yes', and that would seem to be enough to answer the question appropriately. As noted above, to make such a charge stick, the questioner should be held to supporting the allegation by producing evidence of bias or dishonesty.

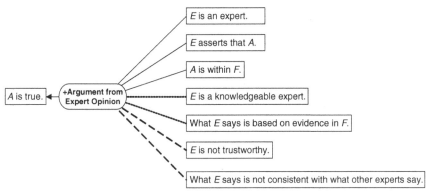

FIGURE 2.9 How Argument from Expert Opinion Is Visually Represented in the Carneades Interface

According to the discussion above, only the consistency and backup evidence questions need some evidence to back them up before the mere asking of the question defeats the original argument. Hence only these two of the critical questions are treated as exceptions. The results of how the critical questions should be classified as premise on the Carneades model can be summed up as follows.

Premise: *E* is an expert.

Premise: *E* asserts that *A*.

Premise: *A* is within *F*.

Assumption: It is assumed to be true that *E* is a knowledgeable expert.

Assumption: It is assumed to be true that what *E* says is based on evidence in field *F*.

Exception: *E* is not trustworthy.

Exception: What *E* asserts is not consistent with what other experts in field *F* say.

Conclusion: *A* is true.

Figure 2.9 shows how argument from expert opinion is visually represented in the Carneades interface. A normal premise is represented by a solid line, an exception is represented by a dashed line, and an assumption is represented by a dotted line. As Figure 2.9 shows, the critical questions are represented as additional premises alongside the ordinary premises in the scheme for argument from expert opinion. This means that, as far as Carneades is concerned, attacking the argument by asking anyone critical questions can be classified as a premise attack argument. According to the Carneades model, the ordinary premises are stated, whereas the other premises expressing critical questions are either assumptions or exceptions.

If we are using Carneades to help us devise a strategy to refute an argument we are confronted with, we can look over the evidence available in the case, or that could possibly be collected in the case, in order to decide which of the critical questions would be the best one to pose. Posing a critical question of the assumption type requires no evidence to back it up in order to defeat the original argument. These would be the first premises to look at. Goodwin described the strategy as one of examining the reasons the other side is giving to support its argument to see if these reasons hold up under critical questioning. However, if there is evidence that could be used to back up one of the critical questions, the backup evidence question would be the one to pose. As we see in the case of Dr. Smith, there is evidence that could be used to back up the claim that he is biased. Hence Carneades can automatically point to the trustworthiness question, represented as an exception in the argument visualization, and indicate that the best strategy is to ask this question.

7. How Carneades Models Attacks and Rebuttals

Not only are schemes classified under other schemes, but critical questions also have a classification structure as well. For example, although argument from bias is a specific type of argument in its own right with its distinctive argumentation scheme, asking a critical question about bias is so common in responding to arguments from expert opinion that it needs to be identified as a specific critical question in its own right with respect to the scheme for argument from expert opinion. In Walton (1997, 213–217) the bias critical question is treated as a subquestion of the trustworthiness question. In other words, questioning whether an expert is biased is treated as a special case of questioning whether the expert is personally reliable as a source. The reason is that questioning on grounds of bias is a way of questioning the trustworthiness of an expert source. A biased expert need not be completely untrustworthy, but if there are grounds for suspecting a bias, that is a good reason for having reservations about the strength or even the acceptability of an argument from expert opinion.

Let's go back to the example Goodwin gave to illustrate the technique of attacking the reasons the other side has put forward in its argument. In this example, the attack alleges that Dr. Smith is biased, because his research is entirely funded by the video game industry. Next, evidence to support this claim of bias is put forward. It is claimed that the 2001 investigation by the Parent's Defense League constitutes evidence to support bias. Let's look back at the pro-contra argumentation displayed in Table 1.3 in a dialogue format. In this example, which we can now see represents part of the argumentation in Goodwin's example above, one party in the dialogue makes the claim that video games do not lead to violence, and supports this claim using an argument from expert opinion attributed to Dr. Smith. The other party then poses any critical question asking whether Dr. Smith could be biased. The first party asks what evidence the other party has for saying that Dr. Smith could be biased. The response given is that his research is funded by the video game industry. So in this example, the respondent can be seen as posing a contra argument against proponent's original argument from expert opinion. The contra argument makes the allegation that Dr. Smith is biased, and then backs up this allegation by offering some evidence to support it. This evidence is the further claim that Dr. Smith's research is funded by the video game industry. But the contra argument goes even further than this. It backs up the claim that Dr. Smith's research is funded by the video game industry by presenting further evidence to back that claim up. This further evidence is provided by the statement that the claim about Dr. Smith's research being funded by the video game industry was shown by a 2001 investigation of the Parent's Defense League.

Finding a better way to model the argumentation and critical questioning in this kind of case led to a new version of Carneades in which refutation is structured differently. In the original version, an exception was modeled as a special kind of premise of an argument, as shown in Figure 2.10. In the new version, an exception is modeled as an undercutter. In other words, the revolutionary change was to accept the device of entanglement within the Carneades method of modeling argumentation. Instead of having arrows go only from text boxes to text boxes, the new version of Carneades allows argument nodes to go to other argument nodes. In particular, it represents the notion of an exception as an undercutter, in virtue of which one argument can attack another. This notion of argument attack makes it different from the notions of argument defeat, refutation and rebuttal. The problem may be that the notion of rebuttal is somewhat ambivalent, in that it appears that it could refer to either one of two kinds of cases. The first one is where one argument attacks another by undercutting it, but does not receive or rebut that argument in the sense of giving a stronger reason to show that the conclusion of the argument is false. So rebuttal could be partly in between. But at any rate, before we discuss the project of giving more precise meanings to these notions of argument attack and rebuttal, let's see how the new version of Carneades represents the structure of the argumentation in Goodwin's example.

In Figure 2.10, we represent a different form of argument opposition where one argument attacks the inferential link of another. This way of displaying the structure of Goodwin's example shows the undercutting argument at the bottom. It is a contra argument attacking the argument from expert opinion above it.

Next we need to look back to Figure 1.7. It was drawn in the Carneades style with a pro argument and contra argument. However, notice that it represents a situation very different from the one represented in Figure 2.10. In Figure 1.7, we had a pro argument leading to the conclusion that

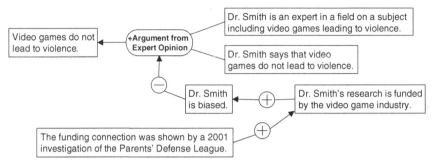

FIGURE 2.10 Argument Map of the Argumentation in Table 1.3

video games do not lead to violence, and we had a contra argument attacking the proposition that video games do not lead to violence. In Figure 2.10, we have an instance of entanglement. First, we have the argument from expert opinion telling us that Dr. Smith is an expert and that since he says that video games do not lead to violence, we can draw the conclusion that video games do not lead to violence. But then we have a counterargument attacking this first argument. According to the counterargument, the claim is made that Dr. Smith is biased, and this claim is backed up by a supporting proposition that is in turn backed up by another proposition. Notice that in this structure the minus node represents a contra argument and the arrow drawn from it leads to the node above that representing a pro argument. In other words, here we have an instance of entanglement.

The argument shown in Figure 1.7 represents an external refutation, we could say, as opposed to the argument map shown in Figure 2.10, which represents an internal refutation. In the type of argumentation represented in the example displayed in Figure 1.7, we had two opposed arguments from expert opinion where the one argument attacked the other. Both arguments in this example were instances of the scheme for argument from expert opinion. In the second example we have only one argument from expert opinion. But this argument is attacked by the asking of the critical question backed up by evidence that strengthens the effect of the critical question is an attack on the prior argument.

The notion of an attack is another concept that needs to be fitted into this system of classification. In the Carneades system, a proposition can be stated, questioned, assumed or accepted. In Carneades one argument can attack another in basically four ways.

1. It can attack one or more of the premises of the prior argument and show that one or more of them is questionable.
2. It can attack one of these premises and show that one or more of them is not acceptable.

3. It can attack the conclusion by posing a counterargument that shows that the conclusion is questionable.
4. It can attack the conclusion by posing a counterargument that shows that the conclusion is not acceptable.

Is an attack the same thing as a rebuttal? At first, it seems that it is, because an attack on an argument is designed to show that the argument is questionable, that it is not supported by the evidence, or even that the evidence shows that it is untenable. On the other hand, it would seem that it is not, because asking a critical question could perhaps be classified as an attack on an argument; it would not seem quite right to say that asking such a critical question is a rebuttal.

This classification may be borderline, however. Asking a critical question casts doubt on an argument, but is casting doubt on an argument rebutting it? What Carneades has shown is that critical questions matching argumentation schemes are of two different kinds in this regard (Walton and Gordon, 2005). Some critical questions act as rebuttals when they are asked, because unless the proponent of the argument replies appropriately to the question, the argument is defeated. Asking other critical questions does not defeat the original argument unless the question is backed up by some evidence. In this kind of case it does not really seem quite right to describe the asking of the critical question as a rebuttal. The word 'rebuttal' also implies that the attacking is being done by posing another argument, and not merely by asking a question about the original argument, even if it is a critical question that casts doubt on the argument.

8. How Carneades Models Relevance

In addition to the three basic ways of attacking an argument listed in Section 1, we also considered some other ways. One of these ways is to argue that the given argument is not relevant to the ultimate conclusion to be proved in the case at issue. To attack an argument in the fourth way, matters of how the argument was used for some purpose in a context of dialogue need to be taken into account. Even though the given argument may stand, having repelled all attacks of the first three kinds, its force as argument may be nullified if it is irrelevant. But is this kind of charge a rebuttal? It is not if it is not an attack on the argument itself, but rather a charge that the argument is not useful for some purpose. A charge of irrelevance is best seen as a procedural objection to the effect that the argument is not useful to resolve the ultimate issue under discussion. To model this kind of procedural objection, we have to look at argumentation as a process, after the manner of Carneades.

The Carneades system can be used to assist an agent preparing a case by constructing arguments used to prove a claim in a situation where there is

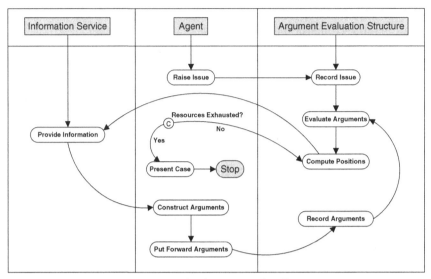

FIGURE 2.11 An Argumentation Process

an information service that continually provides new information that might be useful for this purpose (Ballnat and Gordon, 2010). The agent presents his or her case only once the resources provided by the information service have been exhausted. If that has not happened, the agent tries to make his or her case by asking questions and searching for new information to construct arguments. Then the agent selects which arguments to put forward in order to prove the goal thesis that he or she wants to prove. In this system there is a continuous loop as the agent keeps collecting new information from the information service and uses that information to construct new arguments. A simplified version of this process comparable to the figure in Ballnat and Gordon (2010, 52) is shown in Figure 2.11.

Only once these information and argument construction resources are exhausted does the agent either prove his or her thesis or find that there are insufficient resources to do so. As the agent proceeds through this argumentation process, he or she tries to find alternative positions to support his or her argument.

Suppose I want to prove my claim that proposition *A* is true. What should I do? Should I make a further argument pro *A*? Or should I make another argument con *B*, where *B* is some proposition that is being used by the opposition to refute *A*? Or should I put forward arguments supporting some premise of one of my previous arguments that were put forward in support of *A*? In other words, what should be my next goal, where a goal is a proposition that a party searches for to work on next, by looking for arguments pro or con the proposition he ultimately wants to prove in the dialogue. Carneades is being used here as a device to find which arguments are relevant by telling me which propositions I should choose to work on next, given the information I already have.

As well as providing a method for helping an arguer to determine which arguments are relevant, Carneades can also be used to help an arguer determine which arguments are not relevant. What is presupposed by a claim of relevance is that the given argument is supposed to be used to resolve some unsettled issue in a discussion that is being carried on in the given case. If an argument has no probative value as evidence to prove or disprove the thesis at issue in a particular discussion, it may be dismissed as irrelevant. However, although this attack may knock the argument out of consideration, it is not, strictly speaking, a rebuttal. It should be classified as a procedural objection claiming that the argument under consideration is useless to prove some ultimate claim that the arguer is building a case to prove. On this analysis, the objection to an argument on grounds of relevance is different from the rebuttals and refutations with which we have been concerned. Still, it is interesting to see that Carneades has the capability of dealing with claims of relevance and irrelevance because it can model argumentation as a process.

The procedure recommended for seeking some means of refuting or objecting to an argument broadly follows the line of investigation in the chapter. It starts out by focusing on refutation in the narrower sense, referring to external and internal refutation, then goes on to means of attack and investigation of an argument offered by argumentation schemes and critical questions. From there, it looks more widely to other kinds of objections that may be procedural in nature and that may not focus so narrowly on internal or external refutation. As it expands outward, it takes into account the wider context of an argument, and can do so by viewing argumentation as a process using the Carneades system.

9. Classifying Objections, Rebuttals and Refutations

An objection does not necessarily have to be a counterargument posed against an original argument. It could be merely the asking of a critical question. Even when an objection is a counterargument posed against an original argument, it does not have to be an argument that the original argument is weak, unsupported or incorrect. It could be a procedural objection, not implying that the argument it is addressed against is incorrect, insufficiently supported by evidence or even questionable as an argument in itself. Such a procedural objection could merely claim that the argument, even though it might be reasonable enough, or well enough supported in itself, is not appropriate for use in the context of the given discussion. In law, for example, an argument might be objected to on the grounds that the evidence it purports to bring forward has been obtained illegally, even though that evidence might otherwise be quite convincing in itself as a rational argument. It follows that an objection is not necessarily a rebuttal or a refutation. The term 'objection' represents a wider category.

There is a narrower sense of the word 'objection', however, that is used in logic. Govier (1999, 229) considers an objection to be an argument raised against a prior argument. Hence a question is not an objection: "On this view, a question purely considered as such does not itself constitute an objection". On her account, an objection can be directed in one of two ways. The objection can claim that there is something wrong with the conclusion, or it can claim that there is something wrong with the argument. But these are not the only possibilities. She classifies five types of objections (231), depending on what the objection is specifically raised against: (1) against the conclusion, (2) against the argument in support of the conclusion, (3) against the arguer, (4) against the arguer's qualifications, personal characteristics or circumstances, or (5) against the way the argument or conclusion is expressed. It is interesting to note that some of these categories of objection may correspond to or overlap with types of arguments associated with some of the traditional informal fallacies. The third category and two parts of the fourth may correspond to the *ad hominem* type of argument, while the first part of the fourth may correspond to a common type of attack on arguments from expert opinion.

A different way of classifying objections to an argument has been put forward by Krabbe (2007, 55–57) who lists seven ways an opponent can critically react to a proponent's expressed argument. (1) A request for clarification, explanation or elucidation may contain an implicit criticism that the argument was not clearly expressed to start with. (2) A challenge to an argument comprises an expression of critical doubt about whether a reason supports the argument. (3) A bound challenge raises a more specific doubtful point that offers some reason for entertaining doubt. (4) An exposure of a flaw poses a negative evaluation of an argument and requests further amplification. (5) Rejection is a kind of critical reaction by an opponent who may not deny that the proponent's argument is reasonable, but takes up an opposite point of view. (6) A charge of fallacy criticizes the contribution of the proponent by claiming he or she has violated some rule of fair procedure. (7) A personal attack is a common kind of critical reaction that provides a means of defense against unreasonable moves by one's opponent. Krabbe (2007, 57) suggests that these critical reactions can properly be called objections, because they express dissatisfaction with an argument presented by a proponent. However, Krabbe (2007, 57) writes that to speak of a request for clarification or a pure challenge as an objection would be an overstatement, because objections presuppose a negative evaluation, whereas these other two types of reaction precede evaluation.

There are differences between these two views on what an objection is. Govier (1999, 229) requires that an objection be an argument when she writes, "An objection is an argument, a consideration put forward, alleged to show either that there is something wrong with the conclusion in question or that there is something wrong with the argument put forward in its favor". Krabbe does hold the view that an objection has to be an argument. Ralph Johnson, in an unpublished manuscript shown to the author, has advocated the view that an objection is a response to an argument that can be in the form of a question or a statement and does not have to be an argument. I will take it that objection is a wider category than rebuttal, so that while putting forward a rebuttal is making an objection in some instances, there are also instances in which an objection to an argument should not be classified as a rebuttal.

The notion of a challenge is well known in argumentation. In his Why-Because System with Questions, Hamblin (1970, chapter 8), has a locution 'Why *A*?' that is a challenge or request made to the hearer to provide a justification (an argument) for the statement *A* queried. But what is a challenge to an argument (as opposed to a statement)? Most likely, it would seem to be a critical question. But there could be other sorts of argument challenge; for example, such a challenge could be a procedural objection that the argument is irrelevant.

Following the line of this chapter, the notion of a rebuttal can be defined as follows. A rebuttal requires three things. First, it requires a prior argument that it is directed against. Second, the rebuttal itself is an argument that is directed against this prior argument. Third, it is directed against the prior argument in order to show that it is open to doubt or not acceptable.

A rebuttal is one of a pair of arguments, where the two arguments are ordered, logically rather than temporally, so that the one precedes the other, and so that the second one is directed against the first one. What does "directed against" mean? One argument can have another argument as its target. The one can be meant to support the other or can be meant to attack the other, or the two arguments can be independent of each other. But something more is meant here. What seems to be implied is that a rebuttal is an argument directed against another argument to show that the first argument is somehow defective. To rebut an argument is to try to show that the argument is questionable or not supported by the evidence, or even that the evidence shows that it is untenable.

Is a refutation the same as a rebuttal? One way to define the relationship between these two terms strongly suggested by our discussion of how Carneades handles the type of argument configuration would be to say that a refutation is a successful rebuttal. On this way of defining the two terms, a rebuttal is aimed to show that the argument it is directed against is questionable or untenable. A refutation is a rebuttal that is successful in carrying out its aim. A refutation is a counterargument that not only is posed against a prior argument, but weighs in more strongly when evaluated against the prior argument so that it reverses the conclusion of the prior argument.

So defined, the one term would seem to be a subspecies of the other. A refutation is a species of rebuttal that shows that the argument it is aimed at is untenable. When an argument you have put forward is refuted, it has to be given up. If the argument is confronted with a rebuttal, you may or may not have to give it up. Only if the rebuttal is a refutation do you have to give it up. The same point can be made about attack. Attack does not imply defeat.

The term 'challenge' is widely used in formal dialogue systems. As noted above, Hamblin has a locution, 'Why *A*?', called a challenge in his Why-Because System with Questions. To respond appropriately the hearer is expected to provide premises that the challenger is committed to already or can be brought to concede (in future moves), and *A* is supposed to be a conclusion implied by these premises according to the inference rules in the system. A challenge, in this sense, is not an argument. It is a speech act that requests some evidence to support a claim made by the other party. As the distinction between assumptions and exceptions made in Carneades shows, some critical questions are merely challenges, whereas other critical questions, although they have the speech act format of a challenge, defeat the other party's argument unless he or she comes forward with some evidence to support his or her argument.

The classification tree shown in Figure 2.12 offers a way of clarifying these terms.

Objection is taken to be a wide category that includes procedural objections and many kinds of attacks that should not, strictly speaking, be called rebuttals. An objection of irrelevance is shown as an example of a procedural objection. An objection does not have to be a rebuttal even though it is comparable to a rebuttal in that it assumes that there is something negative about an original argument, or move in argumentation, that needs to be responded to, called into question and corrected. The classification tree in Figure 2.12 incorporates the notion of a challenge. A challenge is defined after the manner of Krabbe as a species of objection that comprises an expression of critical doubt about whether a reason supports the argument that is challenged.

However, this way of defining the notion of challenge makes it appear to be very close to a Pollock-style undercutter, a species of argument attack modeled using entanglement in Carneades. Figure 2.12 clarifies the notion of the challenge by classifying the Pollock-style undercutter as an exception, using the term and its Carneades meaning. Exceptions are classified as critical questions that need to be backed up by evidence before they defeat the argument they are directed against. The classification tree shown in Figure 2.12 also incorporates the distinction between an internal refutation or rebuttal and an external one. Hence it is a comprehensive classification scheme that includes all the species of objections analyzed in the chapter.

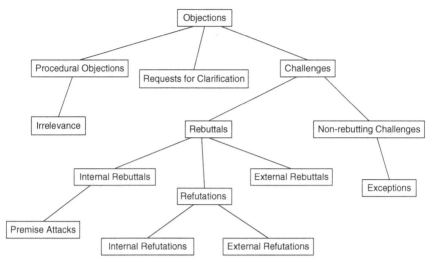

FIGURE 2.12 Classification Tree for Species of Objections

A rebuttal is a species of objection. A refutation is a species of rebuttal that is successful in knocking down the argument it was directed against. A *rebuttal* is an argument directed against another argument to show that the first argument is somehow defective. An *attack*, in the sense of the word as used in the field of argumentation, is an argument directed against another argument to show that the first argument is somehow defective. In other words, for purposes of argumentation study, the words 'rebuttal' and 'attack' can be taken as equivalent.

To rebut an argument is to try to show that the argument is questionable or not supported by the evidence, or even that the evidence shows that it is untenable. A rebuttal can attack a premise of the original argument, it can attack the conclusion, or it can act as an undercutter that attacks the inference from the premises to the conclusion. One way it can do this, as illustrated by Pollock's red light example and the Tweety example, is by finding an exception to a general rule that is the warrant of a defensible argument. A *refutation* is a species of rebuttal that shows that the argument it is aimed at is unacceptable. It could be called a knock-down counterargument. The argument is defeated, and we take the notion of argument refutation as equivalent to the notion of argument defeat. When an argument you have put forward is confronted with a refutation, it has to be given up. Both rebuttals and refutations can be external or internal. It follows that on this view there is a difference between attack and defeat.

The practical argument attack and refutation procedure derived from the analysis in this chapter has seven steps. The procedure can be applied using these seven steps

1. If you have a counterargument that can be used to prove the opposite of the conclusion claimed in the original argument, go for an external refutation.
2. Alternatively, if this seems to be a better route of attack, go for an internal refutation.
3. The first step in seeking a suitable internal refutation is to see if the argument you are trying to attack fits a known argumentation scheme. The list of the most basic types of arguments that have argumentation schemes are the following: argument from position to know, argument from witness testimony, argument from expert opinion, argument from analogy, argument from verbal classification, argument from rule, argument from precedent, practical reasoning, value-based practical reasoning, argument from appearances (perception), argument from ignorance, argument from consequences (positive or negative), argument from popular opinion, argument from commitment, direct *ad hominem* argument (personal attack), circumstantial *ad hominem* argument, argument from bias, argument from correlation to cause, argument from evidence to a hypothesis, abductive reasoning, argument from waste and slippery slope argument.
4. If the argument fits a scheme that can be identified, look at the critical questions matching the scheme and see which question is most appropriate.
5. In the Carneades model critical questions were represented as different kinds of premises, ordinary premises, assumptions and exceptions. Now the exceptions are represented as undercutters, using the device of entanglement.
7. If no part of this procedure so far has come up with a good result, go on to look for some procedural objection, such as questioning whether the argument is irrelevant or circular.

3

Arguments with Missing Parts

This chapter is about how to analyze kinds of arguments traditionally called enthymemes, arguments that require for proper analysis and evaluation the identification of a missing premise, or in some instances a missing conclusion. A small section on enthymemes has traditionally been included in logic textbooks from the time of Aristotle. In this chapter it is shown how methods of argumentation study, including software tools recently developed in computing, have enabled new ways of analyzing such arguments. It is shown how the employment of these methods to four key examples reveals that the traditional doctrine of the enthymeme needs to be radically reconfigured in order to provide a more useful approach to the analysis of incomplete arguments.

There is an extensive literature on incomplete arguments, and this chapter begins with a survey of enough of this literature to make it possible to understand the investigation that follows and to show why it is needed. The second section of the chapter gives a brief historical outline of the literature on enthymemes, beginning with Aristotle's account of it, including coverage of a significant historical controversy about what Aristotle meant by this term. One of the problems with the task of analyzing incomplete arguments is to get some general grasp of what it is one is trying to do, because the solution to this task can be applied not only to logic but to many other fields that contain argumentation, such as science and law. Therefore, it is important to formulate at the beginning what the purpose of the investigation is supposed to be. A brief account of this is contained in Section 3. The next four sections contain extensive analyses of four examples of incomplete arguments. The first example is meant to be very simple, but the next three examples show some highly significant factors found in carrying out the task of argument analysis needed to identify the missing parts of an argument. These findings are taken into account in Section 9 of the chapter, where both the nature of the task and the proper terminology needed to assist it are reformulated and clarified. In this section, a new set of requirements for identifying an argument and its missing parts in a given case is formulated, based on a concept of argument required to support the methods used in the chapter. Section 10 summarizes the conclusions of the chapter.

1. Survey of the Recent Literature on Enthymemes

Some main problems with enthymemes are explained in this section. Ennis (1982, 63–66) drew a distinction between needed and used assumptions. A needed assumption in an argument is a missing proposition such that (1) the argument is not structurally correct as it stands, but (2) when it is inserted, the argument becomes structurally correct (e.g., deductively valid). A used assumption is really meant to be part of the argument by the speaker. Finding used assumptions is much harder, because it depends on what the arguer means. To do this we have to determine, based on the evidence we have, whether the implicit proposition is a commitment of the arguer. This can be a hard task to compute, because we may have to draw inferences from the record of what the arguer said, and since the incomplete argument is stated in natural language, we run up against the usual problems of interpreting the meaning of natural language discourse.

There is an even worse problem (Burke, 1985; Gough and Tindale, 1985; Hitchcock, 1985). If a critic is allowed to fill in an implicit assumption allegedly needed to complete a speaker's argument, he or she may be insert propositions that were not really meant by the proponent to be part of his or her argument. The problem is that the argument analyst seems to be given carte blanche to insert his or her own favorite assumptions and attribute them to the arguer. A check is that he or she should be restrained by the textual evidence visible to all parties, but once again we confront the problem of how to collect and assess that evidence, given that natural language discourse contains ambiguity, vagueness and other phenomena that make this task problematic.

The basic idea of the principle of charity, according to Johnson and Blair (1983, 7), is the obligation of a critic interpreting a particular discourse to treat an argument fairly, "which means to provide the most favorable logical interpretation of that discourse consistent with the evidence". The principle of charity is needed when reconstructing arguments to find premises or conclusions that have not been explicitly stated because there is often more than one candidate proposition that could be used to supply the missing part. It may be hard to make sense out of what the proponent of the incomplete argument was intending to do with it, in some instances. However, the problem with the principle of charity is that it can itself be interpreted in different ways. We would not want to always interpret it in such a strong way as to require the most favorable logical interpretation of an argument that is weak or fallacious by filling in the missing parts so as to make it a valid argument with premises based on solid evidence. In other words, precise requirements for interpreting the discourse in such a way as to treat the argument fairly may not be so easy to formulate.

Lewinski (2008) has provided a nice survey of the literature on the principle of charity, showing how it arose out of the attempts of analytical philosophers to find an answer to the basic question of how to understand each other. In the 1960s and 1970s, philosophers working in the field of informal logic reformulated the principle of charity as a device to help interpret argumentative discourse. According to the classic account of the principle of charity, formulated by Scriven (1976, 71–72), instead of refuting what someone has written by choosing among the various interpretations of the discourse that are possible and choosing the worst one, it can be reasonable to reinterpret the passage in a more charitable way to make more sense out of it. Making sense out of it, Scriven explains (72), is "to make it mean something that a sensible person would be more likely to have meant." According to Scriven, this principle of charitable interpretation has practical value, because when an interpreter chooses a passage that contains some trivial error, it can be easily reformulated to meet the objection of a critic.

Lewinski (2008) also points out, however, that there are problems with the principle of charity. There is even a paradox inherent in it. If an argument analyst is charitable to one party in a discussion where there is a controversy or conflict of opinions involved, it may easily be that the analyst is being uncharitable in representing the views of the other party in the discussion. For any critical discussion, there are two parties who disagree over the issue being discussed, so representing one view in a charitable way then amounts to representing the opposed view in an uncharitable way.

As an alternative to using the principle of charity, which they see as problematic, Paglieri and Woods (2011, 461) analyze incomplete arguments using the notion of parsimony, defined as the tendency to optimize resource consumption in light of an agent's goals. On their analysis, the hearer who receives an incomplete argument from the speaker does not complete it by a cooperative instinct to treat the argument as being reasonable, but by a need to extract valuable information from the message at reasonable cost. Paglieri and Woods (2011, 462) base their analysis of incomplete arguments not on common knowledge, but instead on inferential schemes that enable an argument with missing parts to be completed by the interpreter in a communicative context. Their analysis of enthymemes differs from the traditional one attributed to Aristotle, but preserves the key feature of it, namely, incompleteness. On their view (468), "an enthymeme is an argument in which something essential to its evaluation is not specifically mentioned in its formulation, and thus has to be inferred or known in advance by the hearer".

Walton (2001) showed how enthymemes are often based on implicit premises that can be classified as falling under the heading of common knowledge. Common knowledge has been recognized as important in the literature on argumentation. Govier (1992, 120) categorized a proposition as a matter of common knowledge if it states something known by virtually everyone. She used the examples 'Human beings have hearts' and 'Many millions of civilians have been killed in twentieth-century wars' (120). Freeman (1995, 269) categorized a proposition as common knowledge if many, most, or all people accept it. According to Jackson and Jacobs (1980, 263), in order for rules of conversation to allow participants to engage in collaborative argumentation, there is a need to base many implicit assumptions on commonly shared knowledge. These might be assumptions like, 'Snow is white' or 'Los Angeles is in California'. Common knowledge has also been studied in computing. The open mind common sense system (OMCS) includes statements such as the following ones (Singh, Lin, Mueller et al., 2002, 3) under the category of common knowledge.

- People generally sleep at night.
- If you hold a knife by its blade, then it may cut you.

Common knowledge can be represented in computing by what is called a frame, a data structure for representing a stereotyped situation, like going to a child's birthday party (Minsky, 1975, 2). The power of this theory lies in its inclusion of expectations and other kinds of presumptions.

A frame can be a source of common knowledge used to fill in gaps in an enthymeme. According to Schank and Abelson (1977), common knowledge is based on a *script*, a body of knowledge shared by language users concerning what typically happens in certain kinds of stereotypical situations, and which enables a language user to fill in gaps in inferences not explicitly stated in a text. The research in Walton (2001) did not yield a general solution to the problem of enthymemes, but did analyze several examples of them found in ordinary conversational argumentation, showing that implicit premises based on common knowledge are found in them.

The possibility remains that we might think that we could deal with enthymemes by only using deductive logic, like syllogistic, to fill in missing premises in an incomplete argument.

This possibility has been argued against by van Eemeren and Grootendorst (1984, 127) using the familiar example of the argument that John is English, therefore John is brave. Presumably, the unstated generalization this argument is based on is not the universal one 'All English persons (without exception) are brave' but the defeasible generalization 'English persons are generally (but subject to exceptions) brave'. Paglieri and Woods (2011, 464) used this argument as an example: Ozzie is an ocelot; therefore Ozzie is four-legged. They suggested that this argument rests on the defeasible generalization 'Ocelots are four-legged', on the grounds that Ozzie's failure to be four-legged because of a birth defect would not falsify the generalization.

Walton and Reed (2005) showed how argumentation schemes, representing forms of commonly used defeasible types of arguments, can be applied to an argument found in a text of discourse and used to reveal implicit premises needed to make the argument fit the requirements of the scheme. This method of reconstructing enthymemes was shown to be useful for revealing needed premises in an argument with implicit premises, even though it was conceded that it did not provide an automated enthymeme system that could be mechanically applied to a given argument in a text of discourse to reveal any implicit premises or conclusions in the given argument. It is not hard to see how this method of finding needed assumptions by using defeasible argumentation schemes works. Consider this argument: my doctor says I need vitamin D, therefore I need vitamin D. The missing assumption is that my doctor is an expert in the relevant field (medicine). You can find the missing premise by using the scheme for argument from expert opinion repeated below from the Introduction for the reader's convenience.

Major Premise: Source E is an expert in field F containing proposition A.

Minor Premise: E asserts that proposition A (in field F) is true (false).

Conclusion: A may plausibly be taken to be true (false).

Once the implicit premise 'My doctor is an expert in the relevant field (medicine)' has been inserted into the incomplete argument, the argument fits the requirements of the scheme. Although the example is a simple one, it shows how defeasible argumentation schemes can be used as tools to be applied to a real argument in a natural language of text of discourse to help an argument analyst find the unstated premise or conclusion in that argument. Other schemes include the following ones: argument from witness testimony, defeasible *modus ponens* (DMP), argument from analogy, argument from precedent, practical reasoning, argument from consequences (positive or negative), argument from commitment, argument from correlation to cause and abductive reasoning (inference to the best explanation).

The capability of filling in missing parts of a given argument one is trying to analyze can be provided by computer technology that assists the user to build a visualization of an argument indicating the parts of the argument structure that are based on argumentation schemes. This technology, along with a set of argumentation schemes, shows promise in helping the user to analyze an argument and complete it by filling in missing assumptions. However, as we will see in examples of arguments analyzed later in this chapter, more is required. To analyze examples of arguments with missing parts we will need to be able to distinguish between the commitments of the sender of the argument and those of the receiver to whom it was directed.

As will be shown, this requires viewing an argument as a transaction between sender and receiver. In other words it will require analyzing arguments in a dialogue setting. There are formal dialogue systems for argumentation that are proving to be useful for this purpose.

In the system of Black and Hunter (2008, 439), there are always two agents that act as participants in a dialogue, and each of them takes turns making moves called communicative acts. In their system, a move always takes the form of a triple ⟨*Agent, Act, Content*⟩. For example, the communicative act might be putting forward an assertion. The content of the assertion would be the proposition that is stated in the assertion. Another type of move is asking a question. A dialogue (439) is simply a sequence of moves made from one participant to the other. Each participant in a dialogue has a commitment store that grows over the course of the dialogue as new propositions are inserted into it when a participant makes a move such as putting forward an assertion or an argument.

A dialectical theory of enthymemes (Walton, 2008c) postulated three bases for the enthymeme in a formal dialogue system CBVK: (1) the participants' commitment sets, (2) argumentation schemes shared by both participants and (3) a set of propositions representing common knowledge shared by both participants. The formal dialogue system CBVK is the framework applied to model a notion of implicit commitment used to help analyze incomplete arguments. The main feature of CBVK is its revealing of implicit commitments as unstated premises or conclusions in arguments. CBVK is a formal model of the type of dialogue called persuasion dialogue. In this type of dialogue, there is a conflict of opinions identified at the opening stages of a dialogue, and the goal of the dialogue is for the conflict to be resolved by having an adversarial contest in which each of the two parties brings forward its strongest arguments to support its viewpoint and uses probing criticisms to attack the arguments put forward by the other side. At the closing stage a decision is made, perhaps by an audience or referee that examines and evaluates all the arguments on both sides, to determine which side has the strongest chain of arguments supporting its viewpoint. Persuasion dialogue is partly collaborative, because the participants need to have some base of common knowledge and need to agree on rules for making moves and generally for conducting the disputation, but it is also highly adversarial in nature. The side with the strongest argument wins and the other side loses.

CBVK is built on the previous literature on commitment in dialogue. Walton and Krabbe (1995) built a model of argumentation in dialogue based on Hamblin's (1970) notion of commitment, in which a speaker's commitments do not depend on his or her mental states, but are instead inferred from speech acts that he or she performs in a dialogue. In this sense of the word, a commitment is different from a belief. You can be committed to a statement without believing it is true. You are committed to a proposition when you have gone on record as having asserted it, where assertion is modeled as a speech act in a dialogue. Hamblin (1970; 1971) visualized a commitment store in a dialogue as a set of statements written on a blackboard or stored in a database. As a formal dialogue proceeds, in which parties take turns speaking following protocols governing each move, propositions are added to or retracted from this store, depending on what a speaker does at his or her move. If the proponent asserts a particular statement at a given move, then that statement is automatically inserted into his or her commitment store. While beliefs, desires and intentions are psychological mental states, commitment is a normative notion, defined in a dialectical framework and determined by moves made in a dialogue.

The motivation for studying incomplete arguments of Black and Hunter (2008, 437) is that in order "to build agents that can understand real arguments coming from humans, they need to identify the missing premises with some reliability". The computational model they build for this purpose enables both the proponent and the recipient of an incomplete argument to use the same common knowledge. Their model uses dialogue games made up of communicative acts (speech acts that consist of moves in a dialogue) and protocols, or sets of rules that determine whether or not it is legal to make a move at any given point in a dialogue. The type of dialogue in their study is called an inquiry dialogue, a collaborative type of dialogue in which a group of agents – in the simplest case two – work together to prove a central claim at issue by drawing on a knowledge base that contains a set of propositions representing the evidence in the case. Inquiry dialogues are especially useful in domains such as health care and science that are essentially cooperative in nature. In their model they distinguish between what they call the real argument, the incomplete argument actually presented by the speaker and the intended argument, the completed argument that speaker wishes to communicate to the recipient (Black and Hunter, 2008, 438). The two parties are able to fill in the missing parts and thereby make the transition from the one argument to the other by using common knowledge they both share.

2. Historical Background on the Enthymeme

The word 'enthymeme' derives from the Greek phrase *en thumoi*, meaning 'in the mind'. In modern logic textbooks, an enthymeme is taken to be a syllogism in which one or more of the premises or the conclusion is not stated explicitly but is held in the mind of the arguer. This meaning of the term can be called the traditional doctrine of the enthymeme. To explain why this doctrine has been set forth in so many widely used logic textbooks, such as Copi (1986, 243–247), Burnyeat (1994, 4) offered the answer: "because it was there in the books that Copi read, and for no other (good) reason". In turn, to explain how it came about historically that the traditional doctrine of the enthymeme was so widely entrenched in the logic textbooks, Burnyeat carefully and extensively examined the ancient sources. The main reason is a sentence in the *Prior Analytics* (70a10) cited by Burnyeat (1994, 6): "an enthymeme is an incomplete (*ateles*) *sullogismos* from likelihoods or signs". This statement set in motion the tradition stemming from the earliest commentators on Aristotle's manuscripts telling us that an enthymeme is a syllogism with one or more missing premises. Later on, this view was expanded to include a syllogism with a missing conclusion. A bit of misleading terminology that has helped to confound the issue is that the Greek word *sullogismos*, although it looks like the word 'syllogism', means something different from that, something much broader, even though it could have a general meaning as well as technical meaning in Aristotelian logic (Burnyeat, 1994, 14–15).

When Aristotle's original manuscripts were found, they had to be laboriously transcribed. Some of the significant manuscripts wrote the sentence in question from the *Prior Analytics* without the word *ateles*, while others included it. But this word is omitted from the most significant manuscripts (Tindale, 1999, 10). Another piece of evidence is that there are three different passages in the *Rhetoric* cited by Burnyeat (1994, 8) in which an enthymeme is defined as a *sullogismos* constructed from likelihoods or signs. So we have an interesting conflict here. Did Aristotle mean something different by the term 'enthymeme' when he was writing the *Rhetoric*, or does the meaning expressed in the *Rhetoric* represent the correct definition, meaning that the one given in the *Prior Analytics* is wrong? Burnyeat takes the latter view, and he is not the first one to have done so. There has long been a dissenting view that the traditional doctrine of the enthymeme is untenable. According to Burnyeat (1994, 4), the traditional doctrine of the enthymeme not only is mistaken as an interpretation of Aristotle's writings, but is also "totally useless", even though it is comprehensive and orderly.

Sir William Hamilton (1874, 389–390) argued that the traditional view of the enthymeme is a mistake caused by the later insertion of the word *ateles* into the Aristotelian manuscripts by those who transcribed it, and later by those who wrote commentaries on it. He called the traditional view of the enthymeme a "vulgar doctrine" (1861, 153), arguing instead (1874, 389) that the Aristotelian enthymeme is an argument based on "signs and likelihoods". He argued that arguments from signs are not deductively valid but are instead a kind of inference based on a generalization that something generally appears to be true, subject to exceptions. H.W.B. Joseph (1916, 350) also argued that Aristotle had in mind a species of inference based on a defeasible generalization that holds only for the most part but can be defeated by exceptions. According to Joseph (350), *eikos* is "a general proposition true only for the most part, such as that raw foods are unwholesome." Such eikotic generalizations are subject to exceptions, Joseph contended, and eikotic (enthymematic) inferences based on them hold only tentatively. Joseph (350) cited arguments used in medical diagnosis as examples. Traditionally, such inferences are said to be "probable", but the English word 'probability' (from the Latin *probabilitas*) misleadingly suggests the modern probability calculus. Nowadays we call them defeasible inferences of a kind different from those modeled by the standard Bayesian axioms for probability.

Burnyeat also offered a hypothesis about what the term 'enthymeme' should really be taken to mean in Aristotle's writings. His evidence supporting these claims is comprehensive, and here we can offer only a brief summary of it. He cited numerous examples from Aristotle of the kinds of arguments called Aristotelian topics that represent what he (Burnyeat, 1994, 18) calls a "relaxed" or informal kind of reasoning that is different from the kind of reasoning employed in a deductively valid argument. These informal patterns of argument are familiar to modern argumentation theorists, where they are called defeasible argumentation schemes. Tindale (1999, 11) noted that many of the topics outlined by Aristotle in Book II, chapter 23, of the *Rhetoric* are the same as or similar to the defeasible forms of argument now called argumentation schemes. The advent of argumentation schemes in the literature on argumentation brings to light serious concern that the traditional doctrine of enthymemes is an obstacle to moving forward to developing more precise methods of logical argumentation to identify and analyze arguments found in discourse.

3. A Reorientation of the Problem

What is the purpose of studying incomplete arguments? Is it to improve the argument by making it more persuasive, by making it more logical or by somehow transforming it into a form that is an improvement over the way it was originally expressed in the given text discourse? Or is it a way to set up the argument for critical commentary and evaluation by transforming it into a form where logical tools can be applied to it? These are good questions, but for the present we need to find an entry point by reframing the problem to be addressed. It seems straightforward that since the problem of incomplete arguments traces back to the writings of Aristotle and other Greek philosophers, and has since those times to the present always been treated in logic textbooks, the aim of studying incomplete arguments is a purely logical problem. The problem is one of how to interpret an argument given in a text of discourse in natural language in order to identify the missing parts of the argument that are necessary to use the argument or make sense of it for some communicative purpose. Another purpose is the task of analyzing arguments of the kind found in natural language discourse in order to apply logical tools to them. For example, it is assumed that in teaching logic courses, a good deal of the usefulness of the course lies in its capability of teaching students how to analyze and evaluate particular arguments found in real settings of the kind that is important for the students to deal with. The capability of carrying out such a task rests on the problem of how to find the missing parts of an argument. However, as the examples to be analyzed below will show, the problem has broader implications.

In multiagent systems of the kind currently used in distributed computing, agents have to communicate in order to carry out tasks that require argumentation sequences where questions are asked and arguments are put forward by one agent that need to be understood and responded to by another agent. It would enhance the functionality of such computational argumentation systems if the agents could communicate more economically by not having to state implicit premises and conclusions of their arguments. It would also enhance effective communication and save resources if their arguments could be put forward in a more condensed form by leaving out parts that the other agent could easily insert from components found in the knowledge base that the two agents share.

Still another application is to law where, as indicated by one of the examples treated in this chapter, abbreviated arguments are often put forward, for example, in trials where evidence is being marshaled to prove a disputed claim. In order to understand how such arguments are based on legal reasoning by applying rules to facts, and on commonly used forms of reasoning such as argument from witness testimony and argument from expert opinion, it is often necessary to analyze the arguments carefully to bring out missing assumptions in them. Knowledge about these missing assumptions can be very important to the parties in a trial, for example, to the lawyers to find the weak points in them during cross-examination. Still more broadly, the capability to analyze incomplete arguments could be an important tool in the field of artificial intelligence and law.

As will be shown in the examples of incomplete arguments to be analyzed in the next sections, the so-called problem of enthymemes is by no means as simple as it has been portrayed in the logic textbooks using the traditional approach. In some cases, the missing assumption in a given argument is a premise or conclusion. But in other cases, there is chain of arguments in which the conclusion of one argument also functions as a premise in the next one. In these cases the missing statement can be a premise as well as a conclusion. This complicates matters. But in still other cases, a whole argument is missing that needs to be found and inserted into a tree structure of argumentation where it provides a conclusion or premises needed to connect a chain of argumentation together that has missing parts.

One of the most important tools for undertaking these tasks of argument identification, analysis and evaluation is the argument diagramming method. A beginning point for evaluating any argument found in the text is to construct an argument diagram to represent its premises and conclusions, the inferences that join premises and conclusions, and chaining of arguments representing the sequence of argumentation as a whole. This technique can now be assisted by the use of computational argument mapping tools such as Araucaria, Rationale and Carneades, as shown in Chapter 2. However, to apply any of these extremely useful tools, one has to assume that the argument has already been analyzed, at least to some significant extent, identifying its premises and conclusions. Here we run up against the natural language barrier, because so many of the arguments found in natural language texts are incomplete. To deal with this problem we need to take into account commitments, common knowledge and other factors of how an argument was used for some communicative purpose in a dialogue setting.

For these reasons, on the basis of analyzing four examples in the next four sections, the proposal will be made that it is more useful to replace the doctrine of enthymemes with the doctrine of incomplete arguments. To understand what an incomplete argument is and why it is important, we need to understand something about the context of what we are trying to do in the field of critical argumentation. We are trying to help students of critical argumentation to identify, analyze and evaluate arguments of the kinds found in everyday conversational argumentation, legal argumentation, scientific argumentation and so forth. One of the most common tasks in this field of study is to take a chunk of natural language discourse and identify the arguments in it by using criteria of what constitutes an argument, and from that point onward use other criteria to analyze the argument and to evaluate it as weak or strong. One of the most important tasks of analyzing an argument is to find its missing parts, because arguments are commonly stated in natural language discourse in such a way that not all these parts are explicitly stated. By missing parts, we refer to the propositions that make up the argument, its premises and conclusion. In some instances, one or more premises of the argument may not be explicitly stated. In other instances, the conclusion of the argument may not be explicitly stated. It is widely accepted by those of us who are familiar with and have had long experience with trying to carry out the tasks of argument analysis that is extremely common for arguments to have such missing parts. Hence it is not possible to go forward effectively with the task of evaluating these arguments without having some way of analyzing the argument by identifying these missing parts, even if the second task can be carried out in a professional manner only by constructing hypotheses supported by textual evidence.

4. The Free Animals Example

In this section and the following three sections four examples of arguments with missing parts will be analyzed. It will prove useful to pick a particular system of argument mapping and analysis to represent the structure of argumentation in these examples. Because of its special advantages in

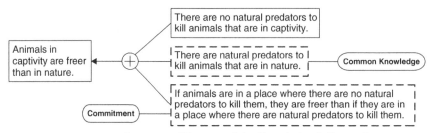

FIGURE 3.1 Argument Diagram of the Free Animals Example

relation to the task of analyzing arguments with missing parts, the Carneades Argumentation System (Gordon and Walton, 2009) has been used.

The Carneades Argumentation System provides a method of visualizing arguments that uses argumentation schemes and applies them to argument construction (invention) as well as to argument analysis and evaluation (Gordon, Prakken and Walton, 2007). Carneades has been implemented using a functional programming language and has a graphical user interface (http://carneades.github.com/). Carneades allows for variations on what happens when a respondent asks a critical question (Walton and Gordon, 2005), and this feature will turn out to be a strong advantage when it comes to analyzing incomplete arguments. It will enable an argument analyst to represent not only the explicit premises of an argument on an argument diagram, but also the critical questions matching an argumentation scheme, which can represent implicit assumptions. Carneades approaches this distinction by distinguishing three types of premises in an argumentation scheme: ordinary premises, assumptions and exceptions. The ordinary premises are the ones explicitly stated as premises of the scheme, but two other kinds of premises are implicit. Assumptions are assumed to be acceptable unless called into question. Exceptions are modeled as premises that are not assumed to be acceptable and that can block or undercut an argument as it proceeds. Ordinary premises of an argument, such as assumptions, are assumed to be acceptable, but they must be supported by further arguments in order to be judged acceptable.

We begin with a simple example that illustrates how missing premises in an argument can be based on common knowledge and on an arguer's commitment. This argument, which we will call the free animals example, was found on a Web site called "Animal Freedom" (http://www.animalfreedom. org/english/opinion/argument/ignoring.html). The argument is: animals in captivity are freer than in nature because there are no natural predators to kill them.

This argument is analyzed in Figure 3.1 using a Carneades argument map in which the conclusion appears in a text box at the far left and the premises are shown at the right as parts of a convergent argument. The plus sign in the node indicates a pro argument with all three premises being parts of the same argument. The explicit premise and conclusion are shown as propositions that appear in ordinary boxes (with solid borders). The two implicit premises appear in the two boxes with the dashed borders.

In the free animals example, the implicit premise that there are natural predators to kill animals that are in nature is classified as common knowledge. The reason is that we all know and accept from common experiences that in nature, animals are constantly killing each other. Cats kill birds, bigger fish eat smaller fish, and so forth. This proposition is not one that needs to be supported by evidence in order for it to be found acceptable.

In contrast let's look at the proposition that if animals are in a place where there are no natural predators to kill them, they are freer than if they are in a place where there are natural predators to kill them. This proposition is controversial, and can easily be subject to doubt or disputation. The source of the example was a Web site called Animal Freedom where controversial arguments about animals and animal rights are put forward and disputed. The conclusion of the argument, the proposition that animals in captivity are freer than in nature, seems paradoxical, or at least questionable, because we normally assume that it is the animals found in nature that are free, whereas the animals in captivity, in a cage, for example, are less free, or perhaps not even free at all. So the claim that animals in captivity are freer than in nature is one that cannot be classified as common knowledge and that appears to represent a special position on the issue of animal freedom that needs to be defended. We don't know the defense of this claim that was offered, but we can suspect that it probably rests on a special meaning of the term 'free' in which an animal might be said not to be free if it is constantly being attacked by other animals trying to kill it. In other words, the proponent of the argument has a particular viewpoint or position on the use of the term 'free' that the audience is trying to get to accept his conclusion that animals in captivity are freer than in nature do not accept as common knowledge. For this reason, on the argument diagram, the proposition that if animals are in a place where there are no natural predators to kill them, they are freer than if they are in a place where there are natural predators to kill them, has been classified as an arguer's commitment. This means that it is being represented as a special commitment of the proponent of this particular argument used to support the conclusion that animals in captivity are freer than in nature. Because it is being represented this way, it is being classified as an argument that needs to be defended in order to be acceptable. And for this reason, it is not being represented as common knowledge. It is not a commitment of both parties to the discussion, but only a commitment of the party who has put forward the argument.

5. The Global Warming Example

The next example is still fairly simple, but introduces a few more complications that are of special interest for the task of analyzing arguments with missing parts. Here is the text of the argument, part of a persuasion dialogue on the issue of global warming: climate scientist Bruce, whose research is not funded by industries that have financial interests at stake, says that it is doubtful that climate change is caused by carbon emissions. The structure of the argument is displayed in Figure 3.2.

In the Carneades-style argument map used to visualize the global warming example in Figure 3.2, the conclusion is contained in the text box at the far left. It is an implicit conclusion of the given argument and is represented in a box with dashed borders. To the right of the conclusion, we see four premises that are parts of an argument from expert opinion. The notation + in the node indicates that this pro argument that has been brought forward to support the conclusion that it is doubtful that climate change is caused by carbon emissions and is based on the four premises that appear to the right of it. The first three premises represent the standard three premises of the argumentation scheme for argument from expert opinion. The implicit premises are the propositions that Bruce is a climate scientist and that Bruce says it is doubtful that climate change is caused by carbon emissions. To make the argument fit this argumentation scheme, however, we have to add the additional assumption that a climate scientist is an expert on climate change. So this additional premise has been added and is contained in a dashed text box. Even though the one in the middle is an implicit premise, all three of these premises are the ordinary premises for the scheme for argument from expert opinion. In this example we can use an argumentation scheme to find an implicit premise.

The next problem is to see where the explicit premise that Bruce's research is not funded by industries that have financial interests at stake fits in. To see this, we have to recognize the function that this premise has in the argument. Its function is proleptic, meaning that its function is to respond to an objection that might be made by the other party, even before the other party has put forward that objection in the conversational exchange (Walton, 2008b). The minus sign on the node indicates this contra argument. In other words, the intended recipient of the argument, the audience, might possibly object to the argument from expert opinion by counterclaiming that the expert, Bruce, is biased, for the reason that Bruce's research is funded by industries that have financial interests at stake.

To see why the argument is analyzed the way it is on the diagram, we have to recall the critical questions matching the scheme for argument from expert opinion (Chapter 1, Section 2).

CQ$_1$: *Expertise Question:* How knowledgeable is E as an expert source?

CQ$_2$: *Field Question:* Is E an expert in the field F that A is in?

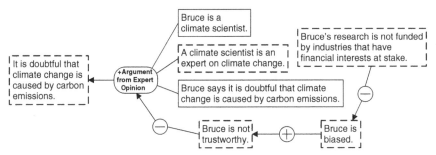

FIGURE 3.2 Argument Diagram for the Global Warming Example

CQ₃: *Opinion Question*: What did *E* assert that implies *A*?

CQ₄: *Trustworthiness Question*: Is *E* personally reliable as a source?

CQ₅: *Consistency Question*: Is *A* consistent with what other experts assert?

CQ₆: *Evidence Question*: Is *E*'s assertion based on evidence?

CQ₁ refers to the expert's level of mastery of the field *F*, whereas CQ₄ refers to the expert's trustworthiness. For example, if the expert has a history of lying or is known to be biased, these findings would undercut the assumption that the expert is trustworthy. The bias critical question is treated as a subquestion of the trustworthiness critical question. In other words, one of the standard reasons for an expert source being classified as not trustworthy is that he or she is biased (e.g., by having something to gain or lose). One way of showing that Bruce is not trustworthy is to show that he is biased. But some evidence of bias has to be given by the respondent in order to make the exception refute the argument from expert opinion.

Looking at the argument diagram, we can see that there is an implicit premise, the statement that Bruce is not trustworthy, represented in a dashed box at the bottom part of the argument from expert opinion. The arrow from the minus node to the node representing the argument from expert opinion indicates that this argument is an exception that undercuts the argument from expert opinion. This means that anyone who wants to cast doubt on the argument from expert opinion by arguing that Bruce is biased has to provide some evidence that Bruce is not trustworthy. Such evidence would imply that the argument from expert opinion is no longer sufficient to support acceptance of the conclusion that it is doubtful that climate change is caused by carbon emissions. Even though this conclusion may have formerly been accepted on the basis that the other three premises of the argument from expert opinion were accepted, now it would no longer be accepted. If the exception were to be supported by evidence, that move would be enough to shift the burden of proof back to the proponent of the conclusion that it is doubtful that climate change is caused by carbon emissions.

So the missing parts of the structure of the argument in this example need to be seen as deriving from not only the commitments of the speaker but also those of the hearer, or audience to whom the argument was directed. The issue is whether climate change is caused by carbon emissions. The proponent of this particular argument is taking one side of the issue. He or she is presenting reasons to support the thesis that it is doubtful that climate change is caused by carbon emissions. The way this issue is framed implies that there is an issue about whether climate change is caused by carbon emissions, and the speaker is trying to persuade those who hold the opposed view to accept his or her view. The audience to whom the argument is directed, we may presume, accepts the opinion that climate change is caused by carbon emissions, or at least is not doubtful about this opinion. Otherwise, there is no need for this argument. The speaker's burden of proof is to persuade this audience that it is doubtful that climate change is caused by carbon emissions.

Once all these contextual assumptions about the supposed purpose of the argument are put in place, we can begin to see how the structure of the argument is based on two additional implicit premises. One standard way to attack the argument would be to argue that since Bruce is biased, he is not trustworthy. We all know that in the climate change dispute one standard way of attacking any argument based on expert opinion to the effect that it is doubtful that climate change is caused by carbon emissions is to claim that the expert source is somehow connected to industries that have financial interests at stake. For example, the scientist's research might have been paid for by corporate interests that have a financial stake in the issue. Since all parties to the argument would be aware of these matters as common knowledge, it would be a good rhetorical strategy for the speaker to rebut this objection proleptically by including a premise argument that can fit into a chain of argumentation that effectively rebuts the exception of his or her argument from expert opinion. In this example, the two implicit premises shown at the bottom of Figure 3.2 are not commitments of the speaker. They are commitments of the other party, who takes the widely accepted view that climate change is caused by carbon emissions. The next example will also involve missing parts based on the commitments of the respondent.

6. The Yogurt Example

The ad called "In Soviet Georgia", designed by the Burson ad agency, was run from 1975 to 1978 on TV and in magazines including *Time* and *Newsweek*. The commercials presented shots of elderly Georgian farmers, and the announcer said, "In Soviet Georgia, where they eat a lot of yogurt, a lot of people live past 100". *Advertising Age* ranked "In Soviet Georgia" as number 89 on its list of the best 100 greatest advertising campaigns. Here is a list of the propositions that make up the premises and conclusions.

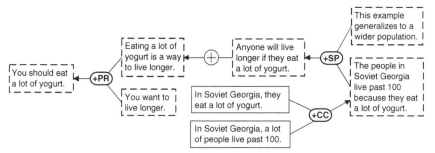

FIGURE 3.3 Basic Argument Diagram 1 of the Yogurt Example

- Explicit Premise: In Soviet Georgia, they eat a lot of yogurt.
- Explicit Premise: In Soviet Georgia, a lot of people live past 100.
- Implicit Premise: The eating of a lot of yogurt is causing the people in Soviet Georgia to live past 100.
- Implicit Conclusion: If you want to live longer, you should eat a lot of yogurt.
- Implicit Premise: You want to live longer.
- Implicit Conclusion: You should eat a lot of yogurt.

An analysis of the structure of the argument is shown in Figure 3.3.

Three argumentation schemes are marked on the diagram: practical reasoning (PR), argument from sample to population (SP) and argument from correlation to cause (CC). The most significant of the three in the analysis given below is the argument from correlation to causation. In this scheme, A and B are variables representing events or kinds of events.

- There is a positive correlation between A and B.
- This correlation is evidence that A causes B.
- Therefore, A causes B.

The notion of positive correlation means that wherever A has been observed, B has also, and the instances in which both occurred together can be counted. To say that A causes B means that A is one of a set of conditions that are (when taken together) sufficient for the occurrence of B, and A is also a necessary condition for the occurrence of B. In addition, A is usually taken to be a condition of a kind that is subject to manipulation. Many instances of arguments that fit this scheme are inherently reasonable, even though they are defeasible and subject to further investigation by the asking of critical questions. However, in some instances the argument can be fallacious. There are eight critical questions matching the scheme.

1. Is there is a positive correlation between A and B?
2. Are there are a significant number of instances of the positive correlation between A and B?

3. Is there good evidence that the causal relationship goes from A to B, and not from B to A?
4. Could there be other causes of B that are more significant than A?
5. Can it be ruled out that the correlation between A and B is accounted for by some third factor C (a common cause) that causes both A and B?
6. If there are intervening variables, then can it be shown that the causal relationship between A and B is indirect (mediated through other causes)?
7. If the correlation fails to hold outside a certain range of causes, then can the limits of this range be clearly indicated?
8. Can it be shown that the increase or change in B is not solely due to the way B is defined, the way entities are classified as belonging to the class of Bs, or changing standards, over time, of the way Bs are defined or classified?

Evaluating an argument from causation to correlation is best carried out in a dialogue format in which the asking of one or more of the critical questions above shifts a burden of proof to the proponent to answer the question, or else he or she has to give up the argument.

An interesting discussion point in this example is whether the argument commits the post hoc fallacy, the error of leaping prematurely from a correlation to a causal conclusion. There are good grounds for concluding, on the analysis above, that the argumentation in this case does commit the post hoc fallacy. The analysis shown in Figure 3.3, along with the scheme and critical questions, provides the right kind of evidence needed to support such a criticism.

The analysis of this case is interesting with respect to the theory of arguments with missing parts because it shows not only an argument with an implicit conclusion, but one with an implicit subconclusion used to link one part of the argument with another. Also, two argumentation schemes can be applied to the structure of the chain of argumentation. We essentially have to chain two arguments connected to each other because an implicit conclusion of the one argument functions as a premise supporting the one premise in the other argument. Even more interestingly, it shows instances of three arguments where the whole argument is composed of missing parts. This observation reveals a whole new aspect of arguments with missing parts. They are not just arguments with missing premises or conclusions. Sometimes, as in this case, a whole chunk of the argumentation containing groups of premises and conclusions is nonexplicit, meaning that whole implicit arguments need to be inserted.

The analysis can be extended further by drilling down to an even deeper level by considering CQ_4 from the list of critical questions matching the scheme for argument from correlation to causation: could there be other causes of B that are more significant than A? This analysis is shown in

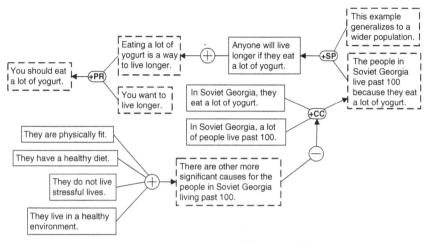

FIGURE 3.4 Deeper Analysis of the Yogurt Example

Figure 3.4, where an exception to the scheme for argument from correlation to cause is shown as an implicit premise in the bottom text box.

The implicit premise in the bottom box of Figure 3.4 is classified as an exception because it has to be supported by evidence by the proponent of the argument if the critical question is asked, or otherwise the argument fails. Hence by analyzing the yogurt example by drilling down to the deeper level shown in Figure 3.4, we have exposed an additional assumption that makes the argument more easily open to criticism. We have critically probed into the argument at a deeper level, once again showing the process of drilling down.

It is also shown in Figure 3.4 how the exception might be supported by additional evidence through argument at the lower left. This ad was successful in the day when people were aware of the longevity of the farmers in Georgia. It was widely thought to be a remarkable phenomenon because there appeared to be no explanation for it. The ad exploited this common knowledge successfully by allowing the reader to jump to an explanation that served the marketers of yogurt products. The same ad would probably not work today, as commonly held opinions about aging and nutrition have changed. Some of the reasons are shown in Figure 3.4 at the bottom right. Awareness of factors such as exercise and nutrition that influence longevity have now become part of common knowledge. This common knowledge makes it much easier for an audience to whom the ad might now be directed to ask critical questions about other possible explanations of the longevity of the people of Soviet Georgia.

The most significant theoretical problem for the analysis of incomplete arguments posed by the yogurt example is that the implicit premises can be classified neither as common knowledge propositions that are accepted by everybody nor as commitments of the proponent of the argument. Instead, they appear to represent commitments of the audience to whom the argument was directed. More accurately, we can presume that they were taken by the proponent of the argument, namely, the advertising agency, to be commitments of the target audience to whom the advertisement was directed. This way of analyzing the argument has a rhetorical aspect, and for that reason it ties in with the concept of the enthymeme found in Aristotle's *Rhetoric*, where the enthymeme is portrayed as a useful device to persuade an audience.

What has been shown by the yogurt example is the lesson that to identify the missing parts of an argument, an analyst needs to recognize different levels of analysis. This is the process of going from a given level to a deeper level of analysis by drilling down. It has been shown how understanding more about this process of drilling down is vitally important to understanding what one is trying to do when one engages in tasks of critical argumentation, especially in the task of finding missing parts of an argument.

7. The Signal Light Example

To begin this section it is necessary to explain a form·of reasoning called defeasible *modus ponens* (DMP). An example from Copi and Cohen (1998, 363) can be used to illustrate DMP: if he has a very good defense lawyer, he will be acquitted; Bob has a very good defense lawyer, therefore he will be acquitted. This argument is defeasible, for even though Bob has a good lawyer, he might not be acquitted. For example, here might be an exception if the lawyer for the prosecution is even better. Using the defeasible conditional symbol =>, the form of DMP can be represented as follow.

Major Premise: $A \Rightarrow B$

Minor Premise: A

Conclusion: B

Many defeasible argumentation schemes have the DMP form. Consider the following version of the argument from expert opinion scheme with an implicit conditional premise added.

Major Premise: Source E is an expert in subject domain S containing proposition A.

Minor Premise: E asserts that proposition A (in domain S) is true (false).

Conditional Premise: If source E is an expert in a subject domain S containing proposition A, and E asserts that proposition A is true (false), then A may plausibly be taken to be true (false).

Conclusion: A may plausibly be taken to be true (false).

More precisely, the argument from expert opinion has the following structure.

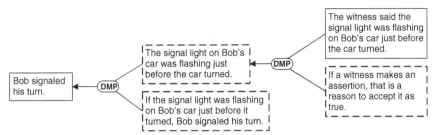

FIGURE 3.5 Argument Map of the Argumentation in the Signal Light Example

Major Premise: (E is an expert and E says that A) => A

Minor premise: E is an expert and E says that A

Conclusion: A

This form of argument is a substitution instance of the DMP form.

Now we have explained the DMP scheme, we can present the fourth example. In contrast to the previous three examples, this fourth one is not a real example found in a text, but instances of it and cases very similar to it are extremely common in legal argumentation, for example, in trials about traffic accidents. It is a good case to illustrate the drilling down technique, where the argument can be represented in a more simple and straightforward way, but then by bringing in more missing parts of it, we can analyze it by exposing a deeper structure of its argumentation. Here is the explicitly stated argument.

- The witness said he saw the signal light flashing on Bob's car just before the car turn.
- Therefore Bob signaled his turn.

The first way of analyzing the argument postulates the following three implicit premises.

- If a witness makes an assertion, that is a reason to accept the assertion as true.
- The signal light was flashing on Bob's car just before the car turned.
- If the red signal light was flashing on Bob's car just before the car turned, Bob signaled his turn.

Connecting the implicit and explicit parts of the argument together using the argumentation scheme for DMP, we get the argument represented in Figure 3.5.

Drilling down, we can identify some additional missing parts.

- The witness was in a position to know whether the signal light on Bob's car was flashing just before the car turned.
- If a witness makes an assertion, that is a reason to accept it as true.

- The signal light on Bob's car was flashing just before the car turned.
- The best explanation of the flashing signal light is that the driver pushed the turn signal indicator.
- Bob was the driver of the car.
- The normal way to signal a turn is for the driver to push the turn signal indicator.

Now the problem is to see how these missing parts fit into the argument. To do this, we have to use two argumentation schemes, the one for abductive reasoning and the one for argument from witness testimony. An abductive inference (Josephson and Josephson, 1994, 14) takes the following form, where H is a variable representing a hypothesis.

D is a collection of data.
H explains D.
No other hypothesis can explain D as well as H does.
Therefore H is plausibly true.

Josephson and Josephson (1994, 14) evaluate abductive reasoning by the following six factors that can be seen as representing critical questions.

1. How decisively H surpasses the alternatives
2. How good H is by itself, independent of considering the alternatives
3. Judgments of the reliability of the data
4. How thorough was the search for alternative explanations was
5. Pragmatic considerations, including the costs of being wrong, and the benefits of being right
6. How strong the need is to come to a conclusion at all, especially considering the possibility of seeking further evidence before deciding.

Abductive reasoning, on this account, is taken to be equivalent to inference to the best explanation (IBE).

Next, we have to introduce the scheme for argument from witness testimony (Walton, 2008d, 60).

Position to Know Premise: Witness W is in a position to know whether A is true or not.

Truth Telling Premise: Witness W is telling the truth (as W knows it).

Statement Premise: Witness W states that A is true (false).

Conclusion: Therefore (defeasibly), A is true (false).

Arguments fitting this scheme can be evaluated using the six critical questions matching the scheme found in (Walton, 2008d, 60).

Using these two schemes, an analysis displaying how the missing parts identified above fit into the argumentation is shown in Figure 3.6. In the analysis of this example visualized in Figure 3.6, eight implicit parts are

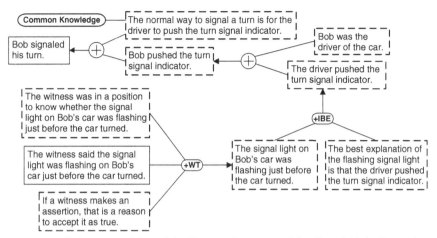

FIGURE 3.6 Argument Map of the Deeper Structure of the Signal Light Example

recognized. Starting with the argument on the bottom left, the argumentation scheme for argument from witness testimony (WT) is applied, revealing two implicit premises and the implicit conclusion that the signal light on Bob's car was flashing just before the car turned. Next, note that both the argument on the right leading to the conclusion that the driver pushed the turn signal indicator and the argument at the top right leading to the conclusion that Bob pushed the turn signal indicator are composed wholly of missing parts. Notice especially that the latter argument has one premise based on common knowledge.

The two arguments along the top of Figure 3.6 are not identified as fitting any argumentation scheme. The method of identifying these arguments is to realize that the sequence of actions represented in them forms a script. Here is the action sequence (script): Bob moved his hand; by moving his hand Bob pushed the turn signal indicator; by pushing the turn signal indicator Bob made the signal light on his car flash; by making the signal light on his car flash Bob signaled his turn. We can see therefore that common knowledge could be exploited in further drilling down into the structure of the missing parts of this example.

Notice that there are no argumentation schemes indicated as applying to the two arguments shown at the top of Figure 3.6. However, we can notice that the implicit premise shown at the top is based on common knowledge about the normal way to signal a turn while driving a car. Figure 3.6 illustrates how this argument combines the two argumentation schemes of argument from witness testimony and abductive reasoning, and combines these with another argument based on a common knowledge premise. In addition, this figure illustrates how three of the arguments in the chain of argumentation are composed of premises and conclusions that were not explicitly stated in the original argument found in the text.

8. Lessons Learned from the Examples

The first problem with analyzing incomplete arguments is the terminological one stemming from the traditional use of the term 'enthymeme' originating from Aristotle and perpetuated in the logic textbooks for over 2000 years. This tradition took the view that an enthymeme is a syllogism with a missing premise. Later the idea of the missing conclusion was added (Burnyeat, 1994), and also the idea of chaining syllogisms together in a sequence of syllogistic reasoning. This way of treating incomplete arguments led to a particular view of the best way to find the missing premise or conclusion. On this view, all you need to do is to focus on a small argument with one premise and one conclusion (or with two premises and a missing conclusion), and then find the missing premise or conclusion and plug it in to complete the argument. The analysis of the last three of the four examples in this chapter has shown that this approach is too narrow. It has been shown here that it is unrealistic and unproductive when applied by an argument analyst to realistic examples of incomplete arguments. It is a view that is an obstacle to finding suitable examples of incomplete arguments by presupposing that such arguments tend to be short syllogistic-like arguments where there is an obvious missing premise or conclusion. The examples analyzed in this chapter, all real or at least realistic examples, showed that incomplete arguments of the kind that are most interesting to analyze are simply not like that.

A problem with enthymemes cited in the literature survey in Section 1 is that if a critic is allowed to fill in any proposition needed to make an incomplete argument valid, he or she may be inserting assumptions that were not meant by the proponent to be part of his argument (Burke, 1985; Gough and Tindale, 1985; Hitchcock, 1985). There is also the even more worrisome danger of committing the straw man fallacy by attributing a premise that distorts the argument in order to make it easier to refute (Scriven, 1976, 85–86). The methods for analyzing incomplete arguments applied in all four examples deals with this problem by showing how the missing parts of a given argument need to be based on the appropriate commitments of the participants in the dialogue. The examples illustrate the requirement that where the missing assumption is taken to be meant by the proponent to be part of his or her argument, it needs to be shown that this missing assumption is acceptable to the proponent, either as common knowledge accepted by all parties to the dialogue or as representing one of the proponent's commitments in the dialogue. This method does not solve the problem of how to analyze and deal with the straw man fallacy, but it does provide a sound theoretical basis for moving ahead with the research on this problem in Chapter 9. The first example, the free animals example, showed how an analyst needs to distinguish between premises that are based on common knowledge and premises that are based on commitment.

The analysis of the last three examples in the chapter has also destroyed another dogma about incomplete arguments, the idea that the commitment that is the missing premise or conclusion is always the commitment of the proponent who is putting forward the argument. Note that in the free animals example, the second implicit premise, as shown in our analysis of the argumentation in that example in Section 4, is classified as a commitment of the proponent who put the argument forward. The proposition 'If animals are in a place where there are no natural predators to kill them, they are freer than if they are in a place where there are natural predators to kill them' was represented as a commitment of the proponent used to support his or her conclusion that animals in captivity are freer than in nature. The reason is contextual. The example came from a Web site called Animal Freedom where debates about animal rights and similar topics are conducted and recorded. The controversy is whether keeping animals in captivity is an ethical practice, and so the type of dialogue is of persuasion dialogue where there is a conflict of opinions on this issue. The proponent's argument cited in the example is arguing for the viewpoint that it is ethically acceptable to keep animals in captivity, and his or her argument represented in Figure 3.1 uses the conclusion that animals in captivity are freer than in nature as one step forward in building a longer argument to support his or her viewpoint as the ultimate conclusion in the chain of argumentation. When posing the question of whose commitment this implicit premise represents, contextual evidence suggests that it is a commitment of the proponent of the argument, the person who put this argument forward.

The situation is more complicated in the global warming example. The implicit premise that a climate scientist is an expert on climate change, we may reasonably presume, is fairly represented as a commitment of the proponent of the argument. However, the other two implicit parts of the argument, the statement that Bruce is not trustworthy and the statement that Bruce is biased, are harder to classify. The arguer's explicit premise that Bruce's research is funded by industries that have financial interests at stake is meant to support the implicit premise that produces bias, which is in turn taken in the next argument to support the conclusion that Bruce is not trustworthy. This latter premise was represented in Figure 3.2 as an exception. What can we say, then, about the two implicit premises that Bruce is biased and Bruce is not trustworthy? Are they commitments of the proponent of the argument, or are they commitments of the respondent or audience to whom the argument was addressed?

It was shown by the analysis of the yogurt example that the implicit premises can be classified neither as common knowledge propositions that are accepted by everybody nor as commitments of the arguer. They need to be analyzed as commitments of the audience to whom the argument was directed. As shown in the analysis of this example, they need to be seen as propositions taken by the advertising agency to be commitments of the target audience. This analysis of the yogurt example needs to be seen as having has a rhetorical dimension. It is about presenting an argument in a way that not only is simple but is based on assumptions that the audience accepts, and for these reasons is persuasive to the audience. This dimension is tied in with the concept of the enthymeme found in Aristotle's *Rhetoric*, where the enthymeme is portrayed as a useful device to persuade an audience (Tindale, 1999).

Finally, the analysis of the signal light example showed the vital importance of analysis of incomplete arguments in legal argumentation. This example illustrates all the features explained above, including common knowledge, use of argumentation schemes and, most important, the technique of drilling down. As shown in Figure 3.6, this technique required the insertion of two entire arguments in order to provide a finer analysis of the argumentation structure of an initial argument that had only a single premise and one conclusion.

9. Refining the Notion of an Argument

To help carry out the task of argument analysis in a useful manner, a precise account of what an argument is taken to be needs to be formulated. The parts of an argument need to be specified, and the requirements for what something has to be in order to constitute an argument need to be explicitly stated. It has been shown by the analyses of the examples in this chapter that the concept of argument required to fit into the procedure of argument analysis useful for providing a method for finding the missing parts of an argument needs to see the concept of argument as having a dual aspect. On the one hand, an argument can be represented as a chain of reasoning visualized in an argument map. On the other hand, an argument needs to be seen as taking place in a context of a verbal exchange in which the claim is being made by one party and in which the claim is subject to doubt by the other party. According to Blair and Johnson (1987, 45), an argument "cannot be properly understood except against the background of the process which produced it – the process of argumentation". This process is initiated "by a question or doubt – some challenge to a proposition" (Blair and Johnson, 1987, 46) and "is a purposive activity" in which each participant has the goal "to change or reinforce the propositional attitude of the interlocutor". Johnson (2000) offered a definition of the concept of an argument that requires an argument to have two basic components: an illiative core and a dialectical tier. In the *illiative core* the reasons supporting a claim are advanced. In the *dialectical tier* known or anticipated objections, alternative positions, criticisms, challenges, questions and reservations are dealt with. The process of argumentation that takes place in the dialectical tier assumes "a minimum of two participants whose roles can be identified as that of questioner and answerer" (Blair and Johnson, 1987, 45). Here we have a concept of argument that is broad enough to work with a method for finding missing parts of an argument. On this view, the illiative core represents the reasons used to support a conclusion, and can include the chain of reasoning connecting premises and conclusions, the kind of structure visualized in an argument map. The dialectical tier represents the notion that the conclusion of an argument is a claim made by one party that is subject to doubt or dispute by a second party, raising critical questions about the argument.

To adapt this dialectical notion of argument to the methods of argument analysis used to identify the missing parts of an argument in this chapter, the concept of argument has to be specified in another respect as well. A distinction needs to be drawn between an argument as an abstract entity and an argument as an item that occurs in some text of discourse, called a *text*. An argument as an abstract entity can in some instances fit an argumentation scheme. Instances of arguments can sometimes be easily identified by people putting them forward or by people hearing or reading them, but not always. In a natural language text there is ambiguity, vagueness and uncertainty about whether something was meant as an argument or not. In addition, as shown, premises or conclusions in an argument found in a text can be unstated.

The initial difficulty faced by the argument analyst is to determine whether the piece of text chosen as the focal point for the analysis really is an argument, as opposed to an explanation or some other speech act. He or she needs to address this task before attempting to find the missing parts needed to complete the argument. But to be able to identify an argument, one has to work with a set of criteria that provides requirements that specify the identifying characteristics of an argument. Moreover, this set of requirements has to be broad enough to incorporate the tools of analysis illustrated in this chapter, including argumentation schemes and the notion of an arguer's commitment to dialogue.

The following set of twelve requirements for identifying an argument and its parts in a given case is tailored to the needs of the task of analyzing an argument to find its missing parts, as illustrated by examples analyzed in the previous sections. Arguments are sometimes hypothetical, but in the normal case we have to deal with in analyzing arguments to find the missing parts of them, the conclusion is a claim being asserted by the proponent of the argument and the premises are meant to provide evidence to support the acceptability of that claim. Hence the following set of requirements reflects this viewpoint.

1. An argument is a set of propositions, some of which are designated as premises, and in the simplest case, one of the propositions is designated as the conclusion to be proved.
2. An argument is an inference from the premises to the conclusion.

3. Arguments can be chained together, so that the conclusion of one argument is also a premise in another. This requirement is the exception to the simplest case cited in requirement 1.
4. An argument is contained in a speech act. In the kinds of cases of incomplete arguments that are the targets of analysis in this chapter, the speech act is that of an assertion, so that the premise and the conclusion are claimed to hold.
5. The conclusion is a claim, that is, an assertion being made by a proponent.
6. The conclusion is subject to doubt by a respondent. To be an argument, as opposed to being an explanation, for example, the speech act has to be directed toward providing evidence to overcome doubt. The purpose of an argument (in this sense) is to prove something.
7. The premises stated by the proponent, and the unstated ones as well, are assumed to hold.
8. An argument has a burden of proof, meaning that if questioned or attacked using a counterargument, it needs to be supported by evidence or the proponent must retract it. There are exceptions, however, for example, a case where all the premises are common knowledge.
9. An argument can be attacked or put in question in three basic ways: (1) by attacking the premise, (2) by attacking the conclusion, or (3) by attacking the inference from the premises to the conclusion.
10. The two parties involved (in the simplest case), the proponent and the respondent, take turns putting forward speech acts that are made in moves in a dialogue.
11. Each move contains a speech act, an action that is made by a participant, and is subject to a response by the respondent (except for the last move in the dialogue).
12. Arguments can have different forms. They can be of different kinds. Some of these kinds are represented by argumentation schemes.

The traditional approach to enthymemes took into account only the first, second and fifth requirements of argument stated above. The limitation of this approach, as shown by the examples studied above, is that it failed to take into account the remaining nine factors. As shown by the analyses in these examples, to cite one factor, the traditional approach failed to take the third requirement into account, the chaining of arguments. The traditional account was working with a definition of 'argument' that is too narrow to permit the kind of analysis of incomplete arguments shown to be required in this chapter. In addition to these twelve basic requirements, it should also be added that arguments have three kinds of premises: ordinary premises, assumptions and exceptions.

As shown in Section 3, the problem of finding the missing parts of an argument has to be seen as part of the task of argument analysis of the kind illustrated in this chapter. This procedure works by applying the abstract structure codified by the twelve requirements to some text that appears to contain an argument. First, the analyst tries to identify the conclusion, then he or she tries to identify the premises being used to support the conclusion. To do this, he or she may have to examine the context in which the argument was being used. For example, an argument used in a newspaper editorial will be on some specific issue being addressed by the author. Knowing something about the issue being addressed can be very helpful in analyzing the argument in the text by determining what its conclusion is supposed to be, judging by the textual evidence.

10. Conclusions

The best terminological hypothesis to move forward with is to abandon the traditional terminology of enthymemes and, instead, base argument analysis on the concept of an incomplete argument, defined as follows. An incomplete argument can be a simple case of a one-step inference that requires for its proper analysis using argumentation methods the insertion of an additional premise or conclusion that was not explicitly stated in the version of the argument given in the text. More complex cases, where there is a chain of inferences (where the conclusion of an argument also functions as a premise in a next one) require the building of an argument diagram, or some comparable tool, along with argumentation schemes, to analyze the structure of the argumentation. Argumentation schemes, along with common knowledge, can be used to fill in missing premises and conclusions in particular arguments that are parts of the wider structure. However, as shown by the examples studied in this chapter, there is often a need to drill down to a finer level of analysis. When this is done, often it is necessary to fill in whole arguments where all the premises and conclusion are implicit assumptions to make the sequence of reasoning in the chain of argumentation hang together properly.

This procedure is shown in Figure 3.7. It starts at the left with the identification of something taken as an argument in a given text. The process of analysis starts by identifying the claim made that is supposedly the conclusion of the argument. This conclusion may be implicit or explicit. If it is implicit, the process of adding the missing parts has begun already. Barring this exception, the procedure carries forward to the next step for the process of analysis to begin. The aim of this process is to identify the parts of the argument more explicitly, and this process can be done by drilling down to different levels. The next step is to add the missing parts, and this process is based on the three components shown at the right side of Figure 3.7: argumentation schemes, common knowledge and commitment in dialogue.

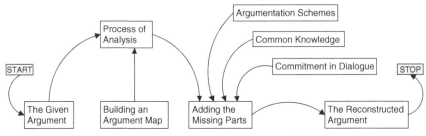

FIGURE 3.7 The Process of Argument Analysis

Once the missing parts have been added, to whatever depth of analysis is needed, the reconstructed argument is then produced as the end point of the procedure.

The antiquated notion of enthymeme is insufficient for the needs of carrying out the task of argument analysis in such cases and, indeed, is even an obstacle to it. Once the dual concept of argument, along with the set of twelve requirements for identifying an argument in a given case, has been adopted as part of the new method, the inadequacy of this traditional notion of the enthymeme is revealed. When undertaking the process of argument analysis, we are not just looking for a missing premise or missing conclusion required to make the argument valid. We are trying to take the given argument available in the text and then using both the evidence of the text in the abstract model of an argument as a normative tool. The aim is to carry through with the process of analysis that produces the reconstructed argument at the other end. In many instances, this process requires building an argument diagram that displays a lengthy sequence of argumentation. In some instances, as shown by the examples in this chapter, several whole implicit arguments may be revealed. By showing how to deploy the concept of an incomplete argument using argument mapping technology of the kind that can now be computationally assisted by software systems, these old prejudices about the enthymeme based on inappropriate ways of defining the basic concepts have fallen by the wayside.

The second main methodological conclusion of the chapter is that the other tool needed for the analysis of these kinds of incomplete arguments is the concept of an arguer's commitment (Hamblin, 1970). The use of this tool requires an approach that takes into account both the reasoning core and the dialectical level. The new method of finding the missing parts of an argument needs to view an argument as more than only a sequence of reasoning of the kind that can be represented on an argument map. It also requires seeing an argument as an orderly back-and-forth exchange between two participants viewed as rational agents that express and possess commitments that can be recorded in a commitment store. It requires an ascent to the dialectical tier.

4

Applying Argumentation Schemes

The aims of this chapter are to survey the resources available for the project of building an exact method that will be helpful for the purpose of identifying arguments in natural language discourse, and to formulate some specific problems that need to be overcome along the way to building the method. It is argued that such a method would be useful as a tool to help students of informal logic identify arguments of the kind they encounter in natural language texts, for example, in newspapers, magazines or on the Internet. The method proposed is based on the use of argumentation schemes representing common types of defeasible arguments (Walton, 1996b; Walton, Reed and Macagno, 2008). The idea is that each scheme is associated with a set of identifiers (key words and markers locating premises and conclusions), and when the right grouping of identifiers is located at some place in a text, the argument mining method locates it as an instance of an argument of some particular, identifiable type (from a list of schemes).

The project is related to the development of argumentation systems in artificial intelligence. One of these technical initiatives, outlined in Section 7, is the project of building an automated argumentation tool for argument mining. The idea is that this tool could go onto the Internet and collect arguments of specifically designated types, for example, argument from expert opinion. These technical initiatives are connected to the aim of finding an exact method for argument identification in informal logic, because the most powerful method would likely turn out to combine both tasks. The most powerful method would have human users apply the automated tool to identify arguments on a tentative basis in a text, and then correct the errors made by the automated tool. It is not hard to see how even a semi-automated procedure of this kind could be extremely helpful for teaching courses in informal logic.

As teachers of logic courses well know, judging whether an argument in a given text of discourse fits some abstract form of reasoning is a sophisticated task with which many beginning students in courses in argumentation and informal logic have recurring problems. Such courses are based on the identification, analysis and evaluation of examples of arguments found in magazines, newspapers and the Internet, or whatever other sources of text materials are available. In order to do an adequate job of teaching an informal logic course, it is necessary to have access to examples of commonly used arguments, and especially types of arguments that tend to be associated with common fallacies, such as arguments from expert opinion, *ad hominem* arguments or appeals to force and threats. As we all know, natural language discourse is full of vagueness and ambiguity, and it can be very hard to pin down a real instance of some text to see whether it fits any abstract structure like a form of argument. Having a procedure for assisting with this task is simply a continuation of the kind of work that is being done every day in teaching courses and writing textbooks in the field of informal logic. However, more exact methods would enable us to find new examples more easily and to document and store them so they could be easily reused.

There are two research initiatives currently under way in argumentation studies that will likely prove to be very helpful to argument analysts confronted with the task of identifying arguments in the natural language text discourse. One is the project of classifying argumentation schemes in a tree structure so that it could be determined how each scheme is related to its neighboring schemes. The other is the project of finding identification conditions for each scheme that could help someone engaged in the task of identifying arguments by providing requirements that a given argument in a text has to meet in order to qualify as fitting a particular scheme. This chapter reports on some findings of this research.

1. Teaching Students of Informal Logic How to Identify Arguments

At the beginning, there are two specific tasks that need to be separated. One is the task of identifying arguments as entities that are distinct from other kinds of entities that occur in natural language discourse, such as explanations. This is the task of distinguishing between arguments and nonarguments. This task is far from trivial, as verbal indicators are often insufficient to distinguish between something that is supposed to be an argument and something that is supposed to be an explanation (van Eemeren, Houtlosser and Snoeck Henkemans, 2007). The other task is that of identifying specific types of arguments. The earlier book on argumentation schemes (Walton, 1996b) identified and described twenty-nine commonly used schemes that represent types of arguments familiar to anyone with a beginner's knowledge of informal logic, as listed below.

1. Argument from analogy
2. Argument from a verbal classification
3. Argument from rule

5. Argument from exception to a rule
6. Argument from precedent
7. Practical reasoning
8. Lack of knowledge arguments
9. Arguments from consequences
10. Fear and danger appeals
11. Arguments from alternatives and opposites
12. Pleas for help and excuses
13. Composition and division arguments
14. Slippery slope arguments
15. Arguments from popular opinion
16. Argument from commitment
17. Arguments from inconsistency
18. Ethotic *ad hominem*
19. Circumstantial *ad hominem*
20. Argument from bias
21. *Ad hominem* strategies to rebut a personal attack
22. Argument from cause to effect
23. Argument from effect to cause
24. Argument from correlation to cause
25. Argument from evidence to a hypothesis
26. Abductive reasoning
27. Argument from position to know
28. Argument from expert opinion
29. Argument from waste

Later work (Walton, Reed and Macagno, 2008) presented a compendium of ninety-six argumentation schemes, depending on how the subtypes are classified.

For example, argument from expert opinion is a common type of argument that we are often interested in for argumentation studies. It is made up of two distinctive premises and a conclusion. Basically, it says: so-and-so is an expert, so-and-so says that some proposition is true, therefore (defeasibly) this proposition is true. Identifying this particular type of argument would seem to be simple. For example, the method could use keywords, such as the word 'expert'. However, from experiences with teaching informal logic methods to students, there is a problem that occurs with some students who will immediately go to the Internet when asked to find examples of this kind of argument and pick the first text they find containing the word 'expert'. Of course, many of these examples are not instances of argument from expert opinion. Keywords that occur in standardized forms of arguments, such as the word 'expert', can be useful in helping a student to find examples of a specific type of argument. But they are crude tools, because their use without further refinement results in many errors.

This kind of work represents a more systematic continuation of the kind of practice that is carried out in teaching courses on argumentation or informal logic. Over many years of teaching courses of this type, I always used basically the same method of starting to teach the students through the use of examples. I searched through magazines, newspapers and the Internet, or whatever other sources of material were available, to find interesting examples of arguments from expert opinion, *ad hominem* arguments, cases of equivocation and so forth. From building up stocks of these cases and discussing and analyzing them with my classes, I started to build up accounts of each of the types of arguments, the kinds of premises they have and the different varieties of them.

However, I did not do this collecting in any systematic way. The examples I found initially came from the news magazine I often read, *Newsweek*, or from the sections on informal logic in the many logic textbooks that use such examples. Eventually, the wealth of experience that came from studying these examples led to the formulation of argumentation schemes, forms used to represent the basic structure of each type of argument. The schemes turned out to be very helpful as I continued to teach courses on argumentation, because they gave students some guideposts to use in their attempts to identify, analyze and evaluate arguments.

There were two kinds of assignments I typically gave to my students in these courses. In the one type, I gave them each the same text of discourse containing an interesting argument, say, a one-page magazine editorial. In the other type of assignment, I asked the students themselves to find an interesting example of one of the arguments we were concerned with in the class, such as argument from expert opinion, and to analyze and evaluate their example. These tasks correspond to what the method built in this project is designed to help with. So it is easily seen how such a method would be helpful for teaching courses of this sort. It would also have a much wider use, however. For example, it would be an extremely powerful tool for researchers in fields such as argumentation and informal logic. They could collect masses of interesting data on particular types of arguments that have long been studied in a more anecdotal way, and make the findings of the field of argumentation study much more powerful, because it would then be based on documented data of a comprehensive sort.

Another example that illustrates how the project will work is the *ad hominem* type of argument. The way I defined this type of argument and crafted the argumentation schemes for it, there has to be more than just a personal attack. For something to be a genuine *ad hominem* argument, four requirements have to be met. First, there have to be two arguers who are engaging in some sort of argumentation with each other. Second, one of the arguers has to put forward an argument. Third, the other arguer has to be attacking the first party's argument. And fourth, the other arguer has to be using personal attack for this purpose. Very often I found that if I asked students to collect an interesting example of an *ad hominem* argument, they would find some instance of name-calling or personal attack, such as "Bob is a liar", and label that as an instance of an *ad hominem* argument. But if the instance of name-calling was not being used to attack somebody's argument, according to the argumentation scheme for the *ad hominem* argument, it should not correctly be so classified. Of course, one can debate the classification system, and there has been plenty of that going on in the field of informal logic, but to carry out a systematic study of any domain, one has to start with some initial hypotheses, definitions and classifications of the things being studied. Hence it is most useful, and in my opinion even necessary, to start with a well-defined set of argumentation schemes, even if the definitions of them are regarded only as tentative hypotheses that are subject to modification and revision as the project processes more and more examples of a given type of argument.

2. Review of Argumentation Schemes

Argument from expert opinion is a subspecies of position to know reasoning, based on the applicability of the assumption that the source is in a position to know because he or she is an expert. In trying to apply these schemes to real cases of argumentation, it can sometimes be easy to get them mixed up. Here is a typical example of argument from position to know.

> If one is trying to find the best way to get to City Hall in an unfamiliar city, it may be helpful to ask a passer-by.
>
> If it looks like this passer-by is familiar with the city, and he or she says that City Hall is 12 blocks east, it would be reasonable to accept the conclusion that City Hall is 12 blocks east.

This form of reasoning is called position to know argumentation.

Where *a* is a source of information, the following argumentation scheme represents the form of position to know argumentation. It is called the scheme for argument from position to know (Walton, Reed and Macagno, 2008, 309).

Major Premise: Source *a* is in position to know about things in a certain subject domain *S* containing proposition *A*.

Minor Premise: *a* asserts that *A* is true (false).

Conclusion: *A* is true (false).

Such an argument can be reasonable in many instances, but it also is defeasible. It can be critically questioned by raising doubts about the truth of either premise or by asking whether *a* is an honest (trustworthy) source of information. The following critical questions match the scheme for the position to know argument.

CQ$_1$: Is *a* in position to know whether *A* is true (false)?

CQ$_2$: Is *a* an honest (trustworthy, reliable) source?

CQ$_3$: Did *a* assert that *A* is true (false)?

The second critical question concerns the credibility of the source. For example, a lawyer cross-examining a witness in a trial is allowed (within controlled limits) to raise critical questions about the character of the witness for honesty. If a witness has been known to lie in previous cases, a cross-examiner is allowed to ask such *ad hominem* questions, as an exception to the general rule against prejudicing the jury, by attacking the ethical character of a defendant.

Let us consider once again the case of asking the passer-by where City Hall is located. Such a case is clearly an instance of position to know reasoning, but is it also an instance of the scheme for argument from expert opinion? Students in a critical thinking course are often inclined to think so, because it may seem to them reasonable to say that the passer-by is being consulted as an expert on the city streets. After all, if he or she is very familiar with them, he or she might be said to have a kind of expert knowledge of them.

The argumentation scheme for argument from expert opinion is different from the one for argument from position to know, because it is required that the source who is in a position to know be an expert. For example, ballistics experts and DNA experts are often used to give expert testimony as evidence in trials, but they must qualify as experts. The basic version of this scheme for argument from expert opinion is given in Chapter 1. It is rarely wise to treat an expert as an infallible source of knowledge, and taking that approach makes argumentation susceptible to the fallacious misuse of argument from expert opinion. As noted in Chapter 1, generally this form of argumentation is best treated as defeasible, subject to failure under critical questioning. The six basic critical questions matching the appeal to expert opinion (Walton, 1997, 223) are listed in Chapter 1. If the respondent asks any one of these six critical questions, the burden of proof shifts back to the proponent's side to respond appropriately.

On this interpretation, the argument would fit the major premise of the scheme for argument from expert opinion. But in the absence of further evidence, can it correctly be said that he or she is an expert? Unless he or she is a cartographer or an expert on city planning, or has some qualification of that sort, he or she would not qualify as an expert in the sense in which the term is used in law. In short, we can draw a distinction between having a working or practical knowledge of some area and having expert knowledge of it.

It might be noted here as well that many arguments that occur in real argumentation texts, whether in law or everyday conversational argumentation, have implicit premises or conclusions (Walton and Reed, 2005). Consider the example "Joao lives in Lisbon and says the weather is fine there, therefore the weather is fine there". An implicit premise is that Joao

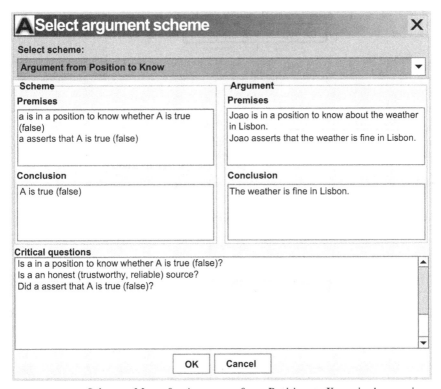

FIGURE 4.1 Schemes Menu for Argument from Position to Know in Araucaria

is in a position to know about the weather in Lisbon, based on the explicit premise that he lives there. Another implicit premise is the defeasible conditional that if a person lives in a place, he or she is in a position to know about the weather there. The implicit conclusion of the argument is the statement that the weather is fine in Lisbon. At present such implicit premises and conclusions can be found only by having a human analyst dig them out as best explanations of the meaning of the text. It should be noted that argumentation schemes are very helpful for this purpose in many instances.

In the Lisbon example, the scheme for argument from position to know can be applied to extract the missing premise and conclusion. Here we use Araucaria (see Chapter 1) to determine that there is an implicit premise in this argument, the proposition that Joao is in a position to know about the weather in Lisbon. We can find this missing premise by selecting the argumentation scheme for 'argument from position to know' shown in Figure 4.1. On the left we see the argumentation scheme for argument from position to know. On the right we see the particular argument about the weather in Lisbon that fits the requirements of this scheme. We also see at the bottom of the menu some critical questions that can be used to respond to an argument fitting the scheme.

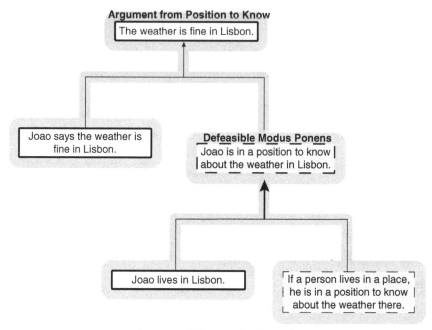

FIGURE 4.2 Argument Diagram for the Lisbon Example

Now we can draw a diagram representing the premises and conclusion of the argument, and the arguments joining them. In Figure 4.2 the missing premise, the statement that Joao is in a position to know about the weather in Lisbon, is inserted in a darkened box with a broken border in the middle of the argument diagram.

As shown in Figure 4.2, the missing premise that Joao is in a position to know about the weather in Lisbon is shown as being supported by two other premises. One is the explicit premise that Joao lives in Lisbon, and the other is the implicit premise that if a person lives in a place, he or she is in a position to know about the weather there. This argument is shown as having the argumentation scheme for defeasible *modus ponens*. *Modus ponens* can take two forms: strict *modus ponens* and defeasible *modus ponens*. Defeasible *modus ponens* has a conditional premise that is open to exceptions. Defeasible *modus ponens* has the following form, where $A \Rightarrow B$ is the defeasible conditional: $A \Rightarrow B$; A; therefore B. For example, if something is a bird, then, generally, subject to exceptions, it flies: Tweety is a bird; therefore Tweety flies. If we find out that Tweety is a penguin, the original defeasible *modus ponens* argument defaults.

Argument from ignorance, or lack-of-evidence reasoning, as it is often called, is another scheme that is so common and natural to use that it is hard to identify. We use it all the time but are scarcely aware we are doing it. This scheme is difficult for students to grasp at first and to identify in natural language discourse, because it is subtle and because it involves negation. The scheme for argument from ignorance is based on both what is known and what is not known to be true at some point in a sequence of argumentation (Walton, 1996a, 254). The major premise is a counterfactual.

Major Premise: If A were true, A would be known to be true.

Minor Premise: A is not known to be true.

Conclusion: A is false.

The major premise is based on the assumption that there has been a search through the knowledge base that would contain A that has supposedly been deep enough so that if A were there, it would be found. The critical questions include considerations of (1) how deep the search has been, and (2) how deep the search needs to be to prove the conclusion that A is false to the required standard of proof in the investigation. In typical instances of the argument from ignorance, the major premise of the argument is not explicitly stated. It has to be extracted from the text by applying the argumentation scheme. It is perhaps also for this reason that students tend to overlook this type of argument and have a hard time identifying it in a given text.

3. Schemes for Practical Reasoning and Arguments from Consequences

There are variants of the scheme for practical reasoning, but the simplest one is the most useful for our purposes here. In the scheme below, the first-person pronoun 'I' represents a rational agent of the kind described by Wooldridge (2000), an entity that has goals, some (though possibly incomplete) knowledge of its circumstances, and the capability of acting to alter those circumstances and to perceive (some of) the consequences of so acting. The simplest form of practical reasoning is the scheme for practical inference introduced in Chapter 1, Section 6.

Major Premise: I have a goal G.

Minor Premise: Carrying out this action A is a means to realize G.

Conclusion: Therefore, I ought (practically speaking) to carry out this action A.

As noted in Chapter 1, Section 6, there are five basic critical questions matching the scheme for practical reasoning (Walton, Reed and Macagno, 2008, 323).

CQ$_1$: What other goals do I have that should be considered that might conflict with G?

CQ$_2$: What alternative actions to my bringing about A that would also bring about G should be considered?

CQ₃: Among bringing about *A* and these alternative actions, which is arguably the most efficient?

CQ₄: What grounds are there for arguing that it is practically possible for me to bring about *A*?

CQ₅: What consequences of my bringing about *A* should also be taken into account?

As noted in Chapter 4, Section 3, the last critical question, CQ₅, is often called the side effects question. It concerns potential negative consequences of a proposed course of action. Just asking about consequences of a course of action being contemplated could be enough to cast an argument based on practical reasoning into doubt.

Another possibility is that an argument based on practical reasoning could be attacked by the respondent claiming that there are negative consequences of the proposed action. This move in argumentation is stronger than merely asking CQ₅, as it is an attempted rebuttal of the original argument. There is a specific argumentation scheme representing this type of argument. Argument from negative consequences cites the consequences of a proposed course of action as a reason against taking that course of action. This type of argument also has a positive form, in which positive consequences of an action are cited as a reason for carrying it out. These are the two basic argumentation schemes for arguments from consequences (Walton, Reed and Macagno, 2008, 332), where *A* represents a state that could be brought about by an agent.

The first is called argument from positive consequences.

Premise: If *A* is brought about, good consequences will plausibly occur.

Conclusion: Therefore *A* should be brought about.

The other scheme of the pair is called argument from negative consequences.

Premise: If *A* is brought about, then bad consequences will occur.

Conclusion: Therefore *A* should not be brought about.

Argumentation from consequences offers a reason to accept a proposal for action tentatively, subject to exceptions that may arise as new circumstances become known. An instance of argument from consequences can be stronger or weaker, depending on its initial plausibility and the critical questions that have been used to attack it.

The scheme for argument from positive value (Walton, Reed and Macagno, 2008, 321) takes the following form.

Premise 1: Value *V* is *positive* as judged by agent *A*.

Premise 2: If *V* is *positive*, it is a reason for *A* to commit to goal *G*.

Conclusion: *V* is a reason for *A* to commit to goal *G*.

The scheme for argument from negative value (Walton, Reed and Macagno, 2008, 321) takes the following form.

Premise 1: Value *V* is *negative* as judged by agent *A*.

Premise 2: If *V* is *negative*, it is a reason for retracting commitment to goal *G*.

Conclusion: *V* is a reason for retracting commitment to goal *G*.

Argument from negative consequences is a form of rebuttal that cites the consequences of a proposed course of action as a reason against taking that course of action.

Another type of argument widely used in the argumentation is the variant of practical reasoning called value-based practical reasoning (Bench-Capon, 2003). The version of this scheme below is from Walton, Reed and Macagno (2008, 324).

Premise 1: I have a goal *G*.

Premise 2: *G* is supported by my set of values, *V*.

Premise 3: Bringing about *A* is necessary (or sufficient) for me to bring about *G*.

Conclusion: Therefore, I should (practically ought to) bring about *A*.

Another version of the scheme for value-based practical reasoning (Atkinson, Bench-Capon and McBurney, 2004, 88) unpacks the notion of a goal into three elements: the state of affairs brought about by the action, the goal (the desired features in that state of affairs) and the value (the reason why those features are desirable).

In the current circumstances *R*
Action *A* should be performed
To bring about new circumstances *S*
Which will realize goal *G*
And promote value *V*.

Note that value-based practical reasoning can be classified as a hybrid scheme that combines argument from values with practical reasoning.

In some cases, it may be hard to identify the type of an argument, because it is not obvious what scheme it fits and/or because some elements of the argument are not explicitly stated. However, in such cases, clues from the context of dialogue can help. Consider the following blood pressure dialogue (Restificar, Ali and McRoy, 1999, 3).

Proponent: Have you had your blood pressure checked?

Respondent: There is no need.

Proponent: Uncontrolled high blood pressure can lead to heart attack, heart failure, stroke or kidney failure.

The respondent's reply, 'There is no need', could be seen as a way of attacking the second premise of the simplest version of the scheme for argument from practical reasoning. He or she is saying, in effect, that having his or her blood pressure checked is not a necessary means to maintain health. This move illustrates the type of rebuttal that is an attack on a premise of an argument. As noted above, the fifth critical question is associated with argument from negative consequence.

It may be possible to reconstruct the proponent's reaction at his or her first move by using this clue. It could be interpreted as an instance of the scheme for practical reasoning if we insert an implicit premise. Restificar, Ali and McRoy (1999, 3) offer no further information about the context of the argumentation in their example, but it seems reasonable to presume that the proponent is concerned about the respondent's health. If so, one could insert as an implicit premise the statement that a goal for the respondent is his or her health. If this is a reasonable assumption, the proponent's argument could be reconstructed as follows.

Implicit Premise: Your goal is to maintain your health.

Explicit Premise: Having your blood pressure checked is a necessary means to maintain your health.

Conclusion: You should have your blood pressure checked.

On this interpretation, the scheme for practical reasoning can help to reconstruct the argumentation sequence, as one can see as follows. The proponent made an argument from practical reasoning. The respondent questioned the major premise of this practical argument. He or she doubts that having one's blood pressure checked is necessary to maintain health. The proponent then provided an argument to support the major premise of his or her practical argument. The above analysis of the example is meant to be simple, for purposes of illustration. A fuller analysis would show how another scheme, value-based practical reasoning, is involved, and also how it is the necessary condition variant of the scheme for practical reasoning that is involved.

4. The Sunk Costs Argument

Argument from waste is a kind of argument where one party is thinking of discontinuing some course of action he or she has been engaging in for some time, and another party argues, "If you stop now, all your previous efforts will be wasted". It is also called the sunk costs argument in economics, where it has traditionally been regarded as a fallacy, even though more recently, it has been thought to be reasonable in many instances. It is typified by the following kind of case (Walton, 2002, 473).

Someone has invested a significant amount of money in a stock or business. Decreasing value and poor performance suggest it might be a good time to pull out and invest the remaining money elsewhere. But because the person has already invested so much in this venture, and would lose so much of it by pulling out now, the person feels that he or she must stay with it rather than take the loss. To abandon the investment would be too much of a waste to bear, given all the money that has been sunk into it at this point. Reasoning on the basis of sunk costs, the person concludes that he or she must stay with this investment, even though the person is convinced that the prospects for its rising in value are not good.

It is not hard to see why the sunk costs argument is often regarded as fallacious in economics and in business decision-making generally, where an investor needs to think of the future, and should not be emotionally tied to previous commitments once circumstances change. However, there are other cases where the sunk costs argument can be reasonable (Walton, 2002), especially those where one's commitments to something one has put a lot of effort into are based on one's values. The sunk costs argument appears to be a species of practical reasoning that is also built on argument from consequences and argument from values. It is a composite argument built from these simpler schemes. So it is a classic case, raising the problem of how this cluster of schemes should be structured in a classification system for argumentation schemes.

The sunk costs argument is a subtype of argument from negative consequences, as can be seen by putting it in the following form.

If you stop doing what you are doing now, that would be a waste.
Waste is a bad thing (negative consequence).
Therefore you should not stop doing what you are doing now.

More precisely, the sunk costs argument has the following argumentation scheme, where A is an outcome of an action and a is an agent (Walton, Reed and Macagno, 2008, 326)

Premise 1: If a stops trying to realize A now, all a's previous efforts to realize A will be wasted.

Premise 2: If all a's previous attempts to realize A are wasted, that would be a bad thing.

Conclusion: Therefore, a ought not to stop trying to realize A.

In Walton (2002) it is shown that the sunk costs argument can be a reasonable form of argument, but also that it is defeasible and open to the following critical questions.

CQ_1: Is bringing about A possible?

CQ_2: If past losses cannot be recouped, should a reassessment of the cost and benefits of trying to bring about A from this point in time be made?

A failure to address a critical question appropriately during changing circumstances, due to an attachment to previous commitments and efforts, is associated with fallacious instances of arguments from sunk costs.

As indicated above, the sunk costs argument needs to be built on argument from negative consequences and argument from negative values. This takes us to the problem of classifying schemes, by building classification trees showing one scheme is a subspecies of another. This problem will be taken up in Section 6.

Another problem we now need to address is of a different sort, although it is related to classification as well. It has to do with recognizing an instance of a scheme in a natural language argument in a given text of discourse. In some instances, a given argument may look like it should be classified as an instance of the scheme for argument from waste, but there may be questions about this classification. An example can be found in an opinion article in the *Western Courier*, October 25, 2008.[1] The article advocates the use of embryonic stem cells for the advancement of medicine and claims that the technology exists for deriving human embryonic stem cells without harming the embryo. The argument opposes the position of conservative groups, who are unwilling to support any kind of embryonic research, regardless of whether or not it destroys the embryo. One of the main arguments appears to be an instance of argument from waste, as indicated in the part of the article quoted below.

This [position] is shortsighted and stubborn. The fact is, fetuses are being aborted whether conservatives like it or not. Post-abortion, the embryos are literally being thrown away when they could be used in life-saving medical research. It has become a matter of religious and personal beliefs, and misguided ones at that. Lives could be saved and vastly improved if only scientists were allowed to use embryos that are otherwise being tossed in the garbage.

The argument in this article could be recast in a format that makes it appear to fit the scheme for argument from waste.

Premise 1: The embryos could be used in lifesaving medical research.

Implicit Premise 1: Lifesaving medical research is a good thing.

Premise 2: The embryos are being thrown away.

Implicit Premise 2: Anything being thrown away that could be used is a waste (a bad thing).

Conclusion: The embryos should be used in medical research.

Putting the argument in this format makes it appear that it is an instance of argument from waste. But is it? This question is puzzling, and opinions on both sides can be found. The word 'waste' is used, and waste is taken to be a

---

[1] Available at www.westerncourier.com/news/2006/09/01/Opinion/Stem-Cells.Are.Going. To.Waste-2255161.shtml.

bad thing. Also, that some action is classified as a "waste" is taken as a reason against it. However, what appears to be missing is that in a proper argument from waste, as required by premise 1 of the scheme, the agent is making some previous efforts to do something, and if he or she stops now, his or her efforts will be wasted. In the stem cell example, there are no previous efforts of this sort. Instead, what is said to be a waste are the embryos that are "thrown away". Premise 2 of the scheme for argument from waste also requires that if previous attempts to realize something are wasted, that would be a bad thing. There seems to be nothing fitting this premise in the stem cells example. Nobody was doing anything with the stem cells previously. No effort or commitment was being put into doing something with them.

The problem with this kind of case is one of matching the premises of the scheme with the premises of the argument given in the text of discourse. If a required premise of the scheme is not found in the argument in the discourse, we need to look and see if there is evidence that it is an implicit premise. If there is no such evidence, we need to conclude that the given argument is not an instance of this particular scheme. In the present example, we need to conclude that the given argument is not an instance of the scheme for the sunk costs argument.

5. Slippery Slope Arguments

Finally, we need to consider the scheme for the slippery slope argument, a highly complex form of argument composed of the other simpler schemes we have so far studied. It is a common problem in teaching critical thinking skills to students that once they are taught a structure for the slippery slope argument, they tend to apply it to cases where the evidence does not really justify classifying it under this category. For example, they typically find cases of argumentation from negative consequences and leap to the conclusion that it must be a slippery slope argument because some bad outcome is being cited as a reason for not carrying out a particular course of action. Strictly speaking, however, to be a slippery slope, an argument has to meet a number of requirements. The following is the scheme for the slippery slope argument given in Walton, Reed and Macagno (2008, 339).

First Step Premise: A_0 is up for consideration as a proposal that seems initially like something that should be brought about.

Recursive Premise: Bringing up A_0 would plausibly lead (in the given circumstances, as far as we know) to A_1, which would in turn plausibly lead to A_2, and so forth, through the sequence $A_2, \ldots, A_n$.

Bad Outcome Premise: A_n is a horrible (disastrous, bad) outcome.

Conclusion: A_0 should not be brought about.

It is an important requirement for this scheme that the recursive premise be present. Without that premise, the argument clearly is simply an instance of argument from negative consequences. It is the presence of the recursive premise that enables us to distinguish in any given case whether the argument is a slippery slope argument as well as being an argument from negative consequences.

The following example is a genuine slippery slope argument. It concerns the burning of an American flag by Gregory Lee Johnson during a political demonstration in Dallas to protest policies of the Reagan administration. Johnson was convicted of "desecration of a venerated object", but the Texas Court of Criminal Appeals reversed the ruling, arguing that Johnson's act was "expressive conduct", protected by the First Amendment. This decision was reaffirmed by the Supreme Court in the case of *Texas v. Johnson* (1989 WL 65231(U.S.), 57 U.S.L.W. 4770). In delivering the opinion of the Court, Justice William Brennan cited the precedent case of *Schacht v. United States*, where it was ruled that an actor could wear a uniform of one of the U.S. armed forces while portraying someone who discredited that armed force by opposing the war in Vietnam.

We perceive no basis on which to hold that the principle underlying our decision in Schacht does not apply to this case. To conclude that the Government may permit designated symbols to be used to communicate only a limited set of messages would be to enter territory having no discernible or defensible boundaries. Could the Government, on this theory, prohibit the burning of state flags? Of copies of the Presidential seal? Of the Constitution? In evaluating these choices under the First Amendment, how would we decide which symbols were sufficiently special to warrant this unique status? To do so, we would be forced to consult our own political preferences, and impose them on the citizenry, in the very way that the First Amendment forbids us to do.

The argument in this text can be identified as an instance of the slippery slope argument. A first step is said to lead to a series of unclear decisions (whether to prohibit the burning of state flags, copies of the Presidential seal, the Constitution, and so forth), which would in turn lead to the outcome of individuals imposing their own political preferences on the citizens. This is said to be an intolerable outcome in a free country, a violation of the First Amendment. In this instance, clearly the recursive premise is present.

The problem with many examples is that the argument does appear to be of the slippery slope type, but the series of intervening steps required to get from the premises to the conclusion is not made explicit. The following excerpt from a letter written by Richard Nixon in the *New York Times* on October 29, 1965, has been taken as an example of a fallacious slippery slope argument in logic textbooks (Walton, 1992, 97), but it is hard to tell whether it really fits this scheme for slippery slope argument. Nixon's letter warned about consequences of the fall of Vietnam in the following terms.

[It] would mean ultimately the destruction of freedom of speech for all men for all time not only in Asia but the United States as well.... We must never forget that if the war in Vietnam is lost ... the right of free speech will be extinguished throughout the world.

This argument certainly looks like a classic case of the slippery slope argument, but where is the recursive premise? The intervening steps are missing. Presumably, what Nixon was claiming is that the fall of Vietnam would lead to the fall of other neighboring countries to Communism, and these events in turn would cause a chain reaction with the final disastrous outcome that the whole world is taken over by undemocratic countries. The problem is that Nixon did not fill in all these intervening steps, and so how can we prove that the recursive premise requirement is really meant by the argument as stated in the example above? One option that needs to be looked at is whether these intervening claims can be taken to be implicit premises. In other words, is the argument an enthymeme? There is evidence for this contention, and thus by marshaling the textual evidence, a case can be made that the argument should properly be classified as a slippery slope. However, the contention that this argument fits the scheme for the slippery slope argument needs to be argued for. If it cannot be sustained by the marshaling of the textual and contextual evidence in the case, the argument should be classified only as an argument from negative consequences that is not also a slippery slope argument.

The slippery slope argument does occur in everyday and legal argumentation, but it is not nearly as common as other schemes mentioned above, such as argument from negative consequences and practical reasoning. As shown by the example above, it is a substantial task to properly identify an argument in a given case as fitting the slippery slope scheme, because the scheme is so complex, with so many prerequisites, and because it is a composite, made up of other simpler schemes. It is argued in Walton (1992) that the slippery slope scheme can be analyzed as a complex chain of subarguments, each having the defeasible *modus ponens* (DMP) structure, but there is insufficient space to discuss this interesting analysis here.

6. Classification of Schemes

The project of automatic identification of arguments in a text using schemes would greatly benefit from a classification system showing which schemes are subschemes of others. The subject of classification schemes is a topic for another book, but it will help to make a few comments on this related project here, since it is so obviously important in the schemes and examples studied above. So far there is no generally accepted system of classification for argumentation schemes. Walton, Reed and Macagno (2008, 349–350) surveyed several different approaches and concluded that it appears that at the present state of development of schemes, the general system summarized below is the easiest to apply. In it, there are three main categories, and various schemes under each one.

REASONING

1. *Deductive Reasoning*
   Deductive *modus ponens*
   Disjunctive syllogism
   Hypothetical syllogism
   Reductio ad absurdum
   Etc.
2. *Inductive Reasoning*
   Argument from a random sample to a population
   Etc.
3. *Practical Reasoning*
   Argument from consequences
   Argument from alternatives
   Argument from waste
   Argument from sunk costs
   Argument from threat
   Argument from danger appeal
4. *Abductive Reasoning*
   Argument from sign
   Argument from evidence to a hypothesis
5. *Causal Reasoning*
   Argument from cause to effect
   Argument from correlation to cause
   Causal slippery slope argument
   (For details, see chapter 5 of Walton, Reed and Macagno, 2008, on causal argumentation).

SOURCE-BASED ARGUMENTS

1. *Arguments from Position to Know*
   Argument from position to know
   Argument from witness testimony
   Argument from expert opinion
   Argument from ignorance
2. *Arguments from Commitment*
   Argument from inconsistent commitment
3. *Arguments Attacking Personal Credibility*
   Arguments from allegation of bias
   Poisoning the well by alleging group bias

Ad hominem arguments
Etc.
4. *Arguments from Popular Acceptance*
Argument from popular opinion
Argument from popular practice
Etc.

APPLYING RULES TO CASES

1. *Arguments Based on Cases*
Argument from example
Argument from analogy
Argument from precedent
2. *Defeasible Rule-Based Arguments*
Argument from an established rule
Argument from an exceptional case
Argument from plea for excuse
3. *Verbal Classification Arguments*
Argument from verbal classification
Argument from vagueness of a verbal classification
4. *Chained Arguments Connecting Rules and Cases*
Argument from gradualism
Precedent slippery slope argument
Slippery slope argument

This classification scheme is very helpful for the purpose of orienting students taking an informal logic course, because it helps group some of the most commonly used schemes into categories. But as shown in the instances of the slippery slope argument and argument from sunk costs, the relationship of these more complex schemes to the simpler schemes that compose them, such as practical reasoning, argument from consequences and argument from values, requires a deeper analysis.

Prakken (2010) has given another example of how schemes are structurally related in an interesting way. He studied the structural relationship between argument from expert opinion and argument from position to know, and showed that the former scheme can be classified as a special instance of the latter scheme. He also showed how argument from evidence to a hypothesis can be analyzed in a manner showing that it is a species of abductive reasoning, often called inference to the best explanation. These findings confirm the hypothesis that many of the most common schemes have an interlocking relationship with other schemes, so that one scheme can be classified as a subspecies of another, but only in a complex manner. This complex manner needs to take into account structural relationships between the schemes.

What has been shown here is very important not only for developing a precise system of classification of schemes, but also for the overarching project of developing a system for argument mining. What has been shown is that there are some simple and basic schemes, and there are some highly complex schemes that are built up as complexes from the simpler schemes. Among the most important simple schemes are practical reasoning, argument from position to know, argument from commitment, and argument from values. Argument from consequences is also a simple scheme, but it has an interesting relationship to the scheme for practical reasoning. Argument from negative consequences corresponds to one of the critical questions matching the scheme for practical reasoning.

7. Research on Argument Mining in Artificial Intelligence

It is encouraging that there are already some systems applying argumentation schemes to legal texts that have been implemented, and the results of this experimental work are very interesting so far (Mochales and Leven, 2009, 27). Discourse theories assume that the structure of a text is that of a graph or a tree and that the elementary units of complex text structures are nonoverlapping spans of text. Moens, Mochales Palau, Boiy and Reed (2007) conducted experiments directed toward the ultimate aim of developing methods for automatically classifying arguments in legal texts in order to make it possible to conveniently access and search types of arguments in such texts. They build on recent work in legal argumentation theory as well as rhetorical structure theory. They look for prominent indicators of rhetorical structure expressed by conjunctions and certain kinds of adverbial groups (2007, 226). They identify words, pairs of successive words, sequences of three successive words, adverbs, verbs and modal auxiliary verbs. Rhetorical structure theory defines twenty-three rhetorical relations that can hold between spans of a text. Most hold between two text spans called a nucleus, the unit most central to the writer's purpose, and a satellite, which stands in a relation to the nucleus. For example, the evidence relation links a nucleus like 'Bob shot Ed' and a satellite like 'Bob's fingerprints were found on the gun'. Their experiments offer an initial assessment of types of features that play a role in identifying legal arguments and single sentences. In future work, they also hope to focus on the classification of different types of arguments.

This work has been applied to legal argumentative texts (Mochales Palau and Moens, 2007; Mochales and Leven, 2009; Mochales Palau and Moens, 2009, 100). In this research the sentences are classified according to argumentation schemes, and the aim is to build a system for automatic detection and classification of arguments in legal cases (Mochales Palau and Moens, 2008). The project has built a corpus from texts of the European Court of Human Rights that was annotated by three annotators under supervision of a legal expert (Mochales and Leven, 2009). This task was made easier by the fact that the court documents that provided their corpus were already classified using subheadings into different parts of the text that had different functions. For example, there is one section of the text where the arguments of the judges are presented. The use of a limited number of argumentation schemes, for example the twenty-six or so identified in Walton (1996b), would be a way to start identifying the different types of arguments. This research opens up opportunities for applying artificial intelligence research to informal logic.

It is interesting to note that there were no identifications of instances of argument from ignorance in the corpus, and very few instances of argument from commitment were identified. Practical reasoning was not used as a scheme in this study. Eighty instances of argument from position to know were found; 2,099 instances of circumstantial argument against the person were found; 10,744 instances of argument from evidence to a hypothesis were found; 2,385 instances of argument from expert opinion were found; 12,229 instances of argument from precedent were found; and 1,772 instances of arguments that fitted no scheme were found.

These results are interesting, but Mochales and Leven (2009, 27) noted a number of problems. To improve the usefulness of systems for automated argument text mining, several research topics are acutely in need of exploration. These observations suggest that what is needed, in addition to more precise definitions of the schemes themselves for use in automated argument detection (Rahwan et al., 2011), is the provision of additional criteria that can be of assistance in determining whether or not a scheme applies to a given argument in a text of discourse in problem cases. Sources for collecting such criteria can already be found in work in artificial intelligence (Moens, Mochales Palau, Boiy and Reed, 2007; Mochales Palau and Moens 2007; Mochales and Leven 2009; Mochales Palau and Moens, 2009) and in argumentation theory (van Eemeren, Houtlosser and Snoeck Henkemans, 2007). The blood pressure example showed that the context of the dialogue in a case needs to be taken into account, as well as the indicator words, in the task of detecting a scheme in discourse. The clue to determining that practical reasoning was the scheme fitting the argument in this case was the critical questioning matching the proponent's use of practical reasoning.

The Amsterdam School has been conducting research for some time on the task of identifying arguments in a text of discourse using so-called argumentative indicators such as 'thus', 'therefore' and 'because' (van Eemeren, Houtlosser and Snoek Henkemans, 2007). A large part of this research has focused on the task of distinguishing between an item in a text of discourse that may properly be taken to represent an argument and some other speech act, such as the putting forward an explanation or making a statement. Only a few argumentation schemes have been studied so far. These include argument from analogy, argument from sign and causal arguments.

Another approach (Wyner and Bench-Capon, 2007) reconstructs legal case-based reasoning in terms of argumentation schemes. This approach uses a set of cases, factors and comparisons between cases to instantiate argumentation schemes from which justifications for an outcome of the case at issue can be derived. These include argument from precedent and argument from analogy. Cases have a plaintiff, a defendant, a set of factors present in the case and an outcome for the plaintiff or defendant (2007, 139). They identify and define what they call the main scheme in a case, including its premises and conclusion (143). One premise of this main scheme is called the factors preference premise, which states that one factor was preferred to another in a previous case decided in the plaintiff's favor. They then introduce a new argumentation scheme they call the preference-from-precedent scheme, which is used to support the factors preference premise of the main scheme. In the general literature on argumentation schemes, this legal scheme would be a particular species of the scheme called argument from precedent (Walton, Reed and Macagno, 2008, 344). They identify other schemes as well, showing how arguments fitting schemes on both sides can be used to support or attack other arguments used in the case at issue. There are some features of their approach that are especially significant. They use an applicability assumption that arises because there might be a number of reasons why an argument put forward in a case is not a suitable precedent for that case. They also distinguish between three different kinds of premises in schemes, called ordinary premises, assumptions and exceptions (Gordon, Prakken and Walton, 2007). This approach is significant because it shows how schemes can be used within the framework of legal case-based reasoning (Ashley, 2006), and especially because it shows how factors can be used to define legal schemes and apply them to argumentation of a legal case.

Rahwan, Banihashemi, Reed et al. (2011) have advanced research on the automated identification of particular schemes by building the first ontology of argumentation schemes in description logic, showing how description logic inference techniques can be used to reason about automatic argument classification. A Web Ontology Language (OWL)-based system is implemented for argumentation support on the Semantic Web. At the highest level, three concepts are identified, called statements, schemes describing arguments made up of statements and authors of statements. Different species of schemes are identified, including rule schemes, which describe the class of arguments, conflicts schemes and preference schemes. The schemes are classified by classifying their components: their ordinary premises, assumptions, exceptions and conclusions. Statements may be classified as declarative or imperative. For example, in the scheme for argument from position to know (see Section 1 above), the class of statement *PositiontoHaveKnowledgeStmnt* is defined as a species of declarative statement associated with the property *formDescription*, 'agent *a* is in a position to know whether statement *A* is true or false', that describes its typical form (Rahwan, Banihashemi, Reed et al., 2011, 8). Using these categories, it is possible to fully describe a scheme, such as the scheme for argument from position to know, by stating the necessary and sufficient conditions for an instance to be classified as falling under this type. Special types of schemes called conflict schemes are identified. The method of identifying schemes is implemented in a Web-based system called Avicenna (2011, 11–13). A user can search arguments on the basis of keywords, structural features and other properties (Rahwan et al., 2011, 12).

It is clear even from this brief description of current work on argument mining in the field of computing that this technical initiative would benefit greatly from more work on refining argumentation schemes. It is also clear that, even though this technical work is only a first step toward the development of useful argument mining technology, there are already interesting implications on how the methods currently being used can be adapted to the needs of informal logic. There are two initiatives that badly need to be carried out to provide resources that could help research both on argumentation technology in artificial intelligence and on informal logic as a subject designed to teach skills of argumentation to logic students or other users. As indicated above, one of these initiatives is to develop a classification system of argumentation schemes. But it is evident from the results of the research summarized above that a top-down approach based on abstract theories of argumentation is by itself not sufficient. What is needed is also a bottom-up approach that examines clusters of schemes that are related to each other and that attempts to determine how schemes in such a cluster are related to each other. By means of this bottom-up approach we can get clear identification criteria that would enable us to disentangle one type of argument from another.

8. Identification Requirements for Types of Arguments

A recent research project at the University of Windsor (Hansen and Walton, 2013) undertook a pilot study to see which kinds of arguments were used by candidates in the recent provincial election in Ontario. During the election, between September and October 2011, some 250 arguments were collected from arguments of candidates found in four leading newspapers. The aim of the study was to find out which kinds of arguments were most commonly used and what some of the characteristics of these arguments were. The study began with a subset of the original (Walton, 1996b) set of schemes, and expanded the subset slightly once it was realized during the course of the investigation that some types of arguments were commonly being used that were not contained in the subset. The study was meant to be a way of testing the completeness of the original subset of schemes in order to help build a more comprehensive set of schemes that would be helpful for the analysis of political argumentation. Only arguments that could be attributed to a candidate directly or indirectly were collected. The argument selected did not include editorials or opinion pieces. The research was directed to determining whether any new schemes should be added to the initial list of schemes chosen for the study. The study also classified types of purposes for which the arguments in the election campaign were used, for example, whether an argument was positive as opposed to being critical, or whether it was critical of another party's policy argument, as opposed to being critical of a person in the other party.

This research project quickly began to encounter the typical problem with working with natural language argumentation discourse. The coders in the project sometimes disagreed about how a particular argument should be classified. For example, should a particular argument be classified as an argument from expert opinion or as an argument from position to know? It was found during the course of the project that one device that proved most useful to help with this problem was the building of a set of identification requirements for each scheme. The identification requirements are designed to help the user judge whether or not the particular argument he or she confronts in the natural language discourse really fits a particular scheme, as determined by the criteria furnished in the identification requirements.

Some examples of such identification requirements for schemes are presented below to give the reader an idea of how they work. After the election project, the set of identification requirements that were used in the project was revised in Walton (2012). The examples presented below are ones used in this revised format. The ones included here are mainly those that have been applied and discussed previously in this book, but a few others are included because of the general interest to readers and because they have features of special interest.

The scheme used most often in this book to illustrate various points is the one for argument from expert opinion.

Identification Requirements for Argument from Expert Opinion

(1) Proposition *A* is subject to doubt and the arguer is looking around for some reason to support the claim that there is some evidence to think that *A* is plausibly true; (2) the arguer has some reason to think that agent *a* has access to evidence on whether *A* is true or not; and (3) his reason for (2) is that *E* is an expert who has special training.

A problem users often have is to be able to classify an argument where there is some doubt about whether it fits the scheme for argument from expert opinion or whether it should be classified as fitting the scheme for argument from position to know.

Identification Requirements for Argument from Position to Know

(1) Proposition *A* is subject to doubt and the arguer is looking around for some reason to support the claim that there is some evidence to think that *A* is plausibly true; (2) the arguer has some reason to think that source *a* has access to evidence on whether *A* is true or not; and (3) the source *a* is not an expert.

Here the scheme for argument from expert opinion is classified as a subscheme under argument from position to know. This decision is an example of bottom-up classification. One can see that the difference between the two schemes is that with the scheme for argument from expert opinion, the source to appeal to has to be an expert, but with the scheme for argument from position to know it is required that the source not be an expert. This approach makes the differentiation simple enough, even though, as experiences from artificial intelligence in the design of expert systems have showed, it is not easy to define concept of expert in a manner that is both precise and has generality. The same problem arises in the use of expert testimony as evidence law. However, both fields have been able to manage to draw this sort of distinction in practice.

Another typical problem with argument mining of natural language discourse is that the argument from lack of knowledge, or argument from ignorance as it is often called, is not easy for beginners to identify. Those of us with advanced skills in argumentation theory can learn to identify it, however, and once we do, we come to realize that it is an extremely common form of argumentation that we use all the time. We use it to close off knowledge base and then draw conclusions about negative evidence, meaning evidence that was not found after a search of the knowledge base. Suppose we want to know whether there is a direct connection between the Minneapolis airport and the Zurich airport, and we look at the listing of all flights at the airport monitor in the Minneapolis airport. We see that no flight is listed between these two cities. We conclude that there is no direct flight between them. This inference is based on argument from lack of knowledge. Even so, beginners have a very hard time recognizing this argument as an instance of this scheme, perhaps partly because it involves negation and partly because drawing an inference from negative evidence is a more subtle form of reasoning than using positive evidence. Below we present the identification conditions for it.

Identification Requirements for Argument from Lack of Evidence

(1) There has to be some knowledge about *A*, but not enough to prove or disprove *A*; (2) there is enough knowledge so that if *A* were true, then it would be known to be true; (3) by default, then, since *A* is not known to be true, it can be assumed to be false.

Another typical problem confronted by users is the difficulty of differentiating between an argument from inconsistent commitments and indirect, or so-called abusive *ad hominem* argument. The identification conditions for the direct *ad hominem* argument make it clear that this is a simple and common form of argument where one party personally attacks another. However, students often identify any kind of personal attack as an *ad hominem* argument, but as the identification conditions make clear below, the argument also has to meet other requirements in order to be fitted into this category.

Identification Requirements for the Direct Ad Hominem Argument

Not only does there have to be (1) an attack on the arguer's character, but (2) this attack has to be used to discredit the arguer's credibility, (3) in order to try to defeat his argument.

Another problem that students often have is to differentiate between argument from inconsistent commitments and the circumstantial type of *ad hominem* argument where the inconsistency is used to discredit another party personally, and that allegation is an attack on the other party's argument. A key expression helping to identify the circumstantial *ad hominem* argument is 'He does not practice what he preaches'.

Identification Requirements for the Circumstantial Ad Hominem Argument

There has to be (1) an attack on the arguer's character, but (2) this attack has to be based on an alleged inconsistency among the arguer's commitments, (3) has to be used to discredit the arguer's credibility and (4) has to be put forward to try to defeat his argument.

The circumstantial *ad hominem* argument is a species of argument from inconsistent commitments. The differentiation between these two types of argument is especially important in studying legal and political argumentation.

Another good case for illustration is the slippery slope type of argument. It is typically, but not always, an instance of argument from negative consequences (Walton, 1992). Students often have a habit of identifying an argument from negative consequences as a slippery slope argument even though it does not meet the other requirements. Not just any argument from negative consequences should qualify as a slippery slope argument. In addition to being an argument from negative consequences, it needs to have a recursive premise and a vague boundary line that produces a so-called gray area (Walton, 1992). Once the arguer enters into the gray area, there is no turning back, often because of the recursive nature of the sequence, and the argument is particularly dangerous because there is no way of knowing exactly when one has entered into the gray area.

Identification Requirements for the Slippery Slope Argument

(1) A decision about an action, or a proposal for action, is being considered; (2) pro and con arguments need to be weighed; (3) there is concern about long-term future consequences the action that has been proposed; (4) there is a gradual sequence from some consequences of others that makes up a lengthy chain of steps; (4) as one proceeds along the sequence of steps, there is a gray area of such a kind that one does not know when one has entered into it; (5) once one has entered into the gray area there is no turning back, and hence the descent to the very bad outcome is inevitable past that point; and (6) there is general agreement that some outcomes are very dangerous or otherwise extremely worthy of avoidance.

One can see that the slippery slope argument is quite a complex form of argumentation, whereas argument from negative consequences is a comparatively simple form of argumentation. It is not hard to see, as well, how the latter form of argument is a species of the former. Another problem, however, is that the slippery slope argument admits of several varieties, and not all of the varieties are instances of the argument from negative consequence. One can see from this example that argument classification for purposes of argument mining is a tricky business.

9. Argument Mining as an Informal Logic Method

How could we use these results to develop argument mining methods that could be used to help students of informal logic identify arguments – for example, arguments of the kind they encounter in natural language texts in newspapers, magazines or on the Internet – and analyze them using standard methods of informal logic? We already have some methods helpful for this sort of task. They include, for example, the use of argument mapping tools to identify the premises and conclusions of arguments and to show how one argument is connected to another in a chain of argumentation. A second, and more ambitious project, is to develop an automated argumentation tool for argument mining. The idea is that this tool could be used on the Internet in the collection of arguments of specifically designated types, such as argument from expert opinion. Of course, these two tasks are connected, because the most powerful method would likely turn out to combine both tasks by having trained human users apply the automated tool to identify arguments on a tentative basis in the text of discourse, and then correct the errors made by the automated tool.

There are six distinct tasks in this endeavor. The first task is the identification of arguments in a text of discourse, as opposed to other entities, such as statements, questions or explanations. Carrying out this task requires some definition of what an argument is, as opposed to the speech acts that can often be confused with putting forward an argument, like offering an explanation. Part of this task is the identification of broad types of argument, such as deductive and inductive arguments, as opposed to the third category, sometimes called plausible arguments. The second task is the identification of specific, known argumentation schemes. The principal way of recognizing a particular argumentation scheme is to be able to identify the premises and the conclusion that make up that scheme. What is required is a parser that can recognize not only the individual units of speech in one of these premises or conclusions, such as nouns and verbs, but also particular nouns occurring in a scheme, such as 'expert', or particular phrases such as 'position to know'. The third task is the deeper classification of argumentation schemes. The fourth task is the more precise formulation of schemes. This can be carried out a number of ways. One way is to formulate schemes that can be applied in a particular field. For example, the scheme for argument from expert opinion needs to be formulated in a more precise way in law than it is for the purposes of analyzing ordinary conversational argumentation, because specific criteria for argumentation from expert opinion as a kind of evidence have already been established in law through legal precedents and court judgments. The fifth task is to develop criteria to enable the differentiation between schemes that appear similar to each other or closely related to each other. The sixth task is to develop techniques for minimizing errors in the identification of schemes in natural language text of discourse. As shown above, in some cases it is easy to confuse one scheme with another. Part of the task here is to develop a corpus of borderline problem cases of this sort, and work on criteria that can be used to solve the problem. An important part of this fifth task is to develop a deeper classification system for argumentation schemes.

There is much work to be done before any useful system of argument identification based on argument mining can be implemented in an informal logic setting. What is needed is to encourage those in the field of informal logic to carry out research projects on the subject. An initial problem for anyone setting up this kind of research project is to decide which kind of natural language texts of discourse should be used as the database. Textbooks in informal logic often take their examples from magazine and newspaper articles, but they sometimes include examples of legal arguments as well. One project would be to take a particular news magazine and try to identify instances of arguments found in it, as well as trying to identify the type of argument. A second project would be to use examples of legal argumentation of some sort. A third project would be to use the database of arguments in Debatepedia, or some similar online source, that contains lots of interesting arguments pro and contra on controversial issues at any given time.

10. Conclusions

The work using schemes for argument mining in legal discourse suggests that in addition to the schemes themselves, additional information that cites specific requirements an argument has to meet to qualify as instance of a particular scheme would be extremely useful. For example, to help tell whether an argument in a text should best be classified under the heading of argument from position to know or argument from expert opinion, some requirements telling the argument annotator what kind of source qualifies as an expert source, like those discussed in Section 1, would be useful. The way to build up such additional resources is to better integrate theoretical research on schemes with the work of testing its application to texts of discourse.

Even though the research work in artificial intelligence offers grounds for optimism about the feasibility of the project of automated argument detection using schemes, the harder problem cases posed above concerning argument from position to know, argument from expert opinion, the sunk costs argument and the slippery slope argument are reasons for concern. It needs to be noted that this task is related to another problem, the problem of enthymemes, or arguments found in a natural language text of discourse that have implicit premises or conclusions. As we saw in tackling the problem of differentiating between slippery slope arguments and arguments from negative consequences, much often depends on implicit premises, like the recursive premise, which the classification of the argument requires but that were not explicitly stated in the given text of discourse. This kind of problem is a central one for argumentation studies. Would it be possible to build an automated system that could detect enthymemes and fill in the missing premises or conclusions so that an analysis of the argument with its missing premises indicated could be provided by an argument visualization tool? The short answer is that it might be a lot more difficult to build such a useful tool of this kind than one might initially think (Walton and Reed, 2005), but the employment of schemes will very definitely be helpful as part of the tool. So the project is worth pursuing for purposes of informal logic, and is closely related to the underlying problem of developing a deeper classification system for argumentation schemes.

5

Similarity, Precedent and Argument from Analogy

This chapter is about the logical structure of argument from analogy and its relationship to legal arguments from classification and precedent. Its main purpose is to provide guidance for researchers in artificial intelligence and law on which argumentation scheme for argument from analogy to use, among the leading candidates that are currently available. Arguments from precedent cases to a case at issue are based on underlying arguments from analogy of a kind extremely common both in everyday conversational argumentation and in legal reasoning. There is a very large literature on argument from analogy in argumentation (Guarini, Butchart, Simard Smith et al., 2009), and the topic is fundamentally important for law because of the centrality of arguments from precedent and analogy in Anglo-American law. It is not hard to appreciate this connection, given that according to the rule of *stare decisis*, the precedent decision of a higher or equal court is binding on a similar current case (Ashley, 1988, 206).

In this chapter, cases are used to argue that arguments from precedent are based on arguments from analogy in legal reasoning and that arguments from analogy are based on a similarity between the two cases held to be analogous. As shown in the chapter, this claim is controversial, because there are different views about how the argumentation scheme for argument from analogy should be formulated (Macagno and Walton, 2009). According to the version of the scheme for argument from analogy argued to be the basic one in this chapter, one of the premises has a requirement holding that there is a similarity between the two cases in point. In this chapter I show how to analyze this notion of similarity using the story-based approach of Bex (2011) and the formal dialogue model for investigating stories of Bex and Prakken (2010). It is shown how an abstract structure called a story scheme can be employed in a way that makes it useful to identify, analyze and evaluate arguments from analogy, and show their function in case-based reasoning where precedents are involved.

In *Popov v. Hayashi* (WL 31833731, Cal. Superior, December 18, 2002), a case that has become a benchmark in artificial intelligence and law (Gordon and Walton, 2006b; Wyner, Bench-Capon and Atkinson, 2007), the issue concerned which fan had ownership rights to a home run baseball hit into the stands by Barry Bonds, while the precedent cases concerned the hunting and fishing of wild animals. A problem posed is that the baseball case and the animal cases don't seem all that similar to each other at first sight, even though it can be argued that they are similar (or not) in certain respects. The problem is to specify exactly how they are similar, or are supposed to be, in an argument from a precedent case to a case being decided, when the relationship between the two cases is thought to be one of similarity. Ashley (2009, 1), referring to one of the animal cases, posed the problem in the question: "How is Barry Bonds' 73rd home run like a fox in a fox hunt?" The problem is to clearly define similarity in such a way that it can identified as being claimed to hold in a pair of cases, so that it can be used as a premise in an argument from analogy. This problem is not so easy to solve as it may initially appear to be, for as Ashley (2009, 1) observed, in legal argument from analogy it is often necessary to interpret similarity and difference at multiple levels.

1. The Wild Animal Cases and the Baseball Case

In the case of *Popov v. Hayashi*, a valuable home run ball was hit into the stands by Barry Bonds in 2001, and a dispute arose concerning which fan had ownership rights to it. In the trial, the reasoning partly turned on some precedent cases that concerned the hunting and fishing of wild animals. Much has been written in the literature on artificial intelligence and law, on its relationship to these other cases and how case-based reasoning can evaluate the argumentation in them using factors and dimensions in analogous cases (Bench-Capon, 2009; 2012).

The following account of the facts of the baseball case has been summarized from the statement of decision of the judge, Kevin M. McCarthy (McCarthy, 2002). Barry Bonds hit his record-breaking seventy-third home run in 2001 at PacBell Park in San Francisco. The ball would be very valuable; Mark McGwire's seventieth home run ball hit in 1998 sold for $3 million. This time the ball went into the stands and landed in the upper portion of the webbing of a glove worn by a fan, Alex Popov. The glove stopped the trajectory of the ball, but the ball did not go fully into the mitt. The partial catch did not give certainty of obtaining control of the ball, since Popov had to reach for it and may have lost his balance while doing this. Just as it entered his glove, he was thrown to the ground by a mob of fans who were also trying to get the ball. Buried face down on the ground under several layers of people, he was grabbed, hit and kicked. Somebody in the crowd videotaped the incident. Another fan standing nearby, Patrick Hayashi, picked up the loose ball and put it in his pocket. When the man making the videotape pointed the camera at Hayashi, he held the ball in the air for the others to see. Hayashi was not part of the mob that had knocked Popov down and was not at fault for the assault on Popov.

According to a tacit code of conduct concerning baseball fans' understanding of first possession of baseballs (Gray, 2002, 6), a fan who catches a ball that leaves the field of play has the right to keep the baseball. However, a fan who tries to catch such a ball but does not complete the catch has no rights to the baseball. The catch occurs only when the fan has the ball in his hands or glove and the ball remains there after its momentum has ceased, and after the fan makes contact with a railing, a wall, the ground or other fans who are trying to catch it. If no one catches the baseball, another fan may pick it up and thereby becomes the owner of it. According to these rules, it looks like Hayashi had the right to ownership of the ball, but Popov took the case to court to contest this claim.

The fundamental disagreement in the trial in the Superior Court of California City and County of San Francisco was about the definition of possession (McCarthy, 2002, 5). In order to aid the court, Judge McCarthy asked four distinguished law professors to participate in a forum to discuss the legal definition of possession. The professors disagreed, and Judge McCarthy admitted that although the term 'possession' appears throughout the law, its definition varies, depending on the context in which it is used. The task of the court was taken to be to craft a definition of 'possession' that applies to the circumstances of the case (McCarthy, 2002, 6). Professor Brian E. Gray was one of the legal experts asked to provide advice, and Judge McCarthy adopted as his central tenet what he called Gray's Rule, the rule that to have possession of the ball, the actor must retain control of it after incidental contact with people and things (McCarthy, 2002, 8). Judge McCarthy (2002, 9) ruled that although Popov did not retain control of the ball, other factors need to be considered. One is that his efforts to retain control were interrupted by a violent mob of wrongdoers. Another is the principle that if an actor takes steps to achieve possession of a piece of abandoned property, but is interrupted by the actions of others, he has a pre-possessory interest in the property. After examining all the arguments, Judge McCarthy decided that any award to one party would be unfair to the other, and that each had an equal and undivided interest in the ball.

During their testimony, the law professors pointed out several precedent cases where there was pursuit of an animal that the pursuer failed to catch because somebody or something intervened, and the issue was whether the pursuer could claim possession of the animal. In *Pierson v. Post* (3Cai. R. 175; 1805 N.Y. LEXIS 311), Pierson was out with hounds chasing a fox when Post captured and killed the fox, even though he knew it was being pursued. The court decided in favor of Post on the grounds that mere pursuit did not give Pierson a right to the fox as his property. In *Young v. Hitchens* (6 Q.B.606 (1844)), Young was a commercial fisherman who spread his net, and when it was almost closed, Hitchens went through the gap and caught the fish with his own net. The court found for Hitchens. In *Keeble v. Hickeringill* ((1707) 103 ER 1127), P owned a pond and made his living by luring wild ducks there with decoys, shooting them and selling them for food. Out of malice, D used guns to scare the ducks away from the pond. In this case P won. In *Ghen v. Rich* (8 F.159 D. Mass, 1881), Ghen harpooned a whale from his ship and it was washed ashore. It was found by another man, who sold it to Rich. According to custom, the man who found the whale should have reported it to Ghen and collected a fee. The court found for Ghen. Gray (2002) cited a number of comparable cases from whaling where possession was defined by taking the accepted customs and practices of the whalers into account.

What makes these wild animal cases work as precedents that can be taken into account in the Popov case, and suggest a conclusion that ought to be drawn favoring one side or the other? An obvious and widely accepted answer is that the animal cases are similar to the Popov case. But what does this answer amount to? On the surface, the cases are not similar. Gray (2002, 1) made the point that catching a baseball is not similar to mortally wounding a fox or harpooning a whale: "a baseball at the end of its arc of descent is not at all like a fox racing across the commons, acting under its own volition, desperately attempting to evade death at the hands of its pursuers". At first sight, the two kinds of cases do not appear to be similar. They are about very different activities. Evidently, the similarity is only that they are both about one party trying to catch and possess something, and about interference by another party who also seeks possession of the same thing in a way that might prevent another from obtaining possession. That's not what we normally think about when we say that two things are similar. We think of them sharing a lot of properties of a visible kind so that they look similar. In law, however, features such as intentionality may need to be taken into account.

2. Arguments from Analogy and Precedent in Law

The literature on argument from analogy in fields spanning logic, argumentation studies, computer science and law is enormous. Many proposals have been put forward to represent argument from analogy as a form of reasoning or argumentation scheme, and there is no space to try to summarize them here. We can only refer the reader to the summary of some of the leading theories in (Macagno and Walton, 2009) and the multidisciplinary bibliography of Guarini, Butchart, Simard Smith et al. (2009). Instead, we concentrate on two particular proposals to represent the structure of this argumentation scheme that provide a useful contrast to focus the discussion.

The simplest argumentation scheme for argument from analogy can be represented by this first version from Walton, Reed and Macagno (2008, 315).

Similarity Premise: Generally, case C_1 is similar to case C_2.

Base Premise: A is true (false) in case C_1.

Conclusion: A is true (false) in case C_2.

Let's call this scheme the basic scheme for argument from analogy. The assumption behind the basic scheme for argument from analogy is that similarity between two cases where A holds in the one case can shift a weight of evidence to make plausible the claim that A also holds in the other case. This kind of argument is defeasible, and it can in some instances even be misleading and fallacious, as the traditions of informal fallacies warn us (Hamblin, 1970). But how can similarity be defined or measured? It seems at first that it can be defined in visual terms as an overall appearance of likeness perceived between two cases. It is an important kind of argument to study, because so much of our reasoning is based on it (Schauer, 2009). This kind of similarity is so striking in some instances, at least at first impression, that it makes the person to whom the argument is directed ignore other relevant evidence.

It doesn't seem to be this type of argument from analogy, however, that is being employed in the arguments from precedent from the animal cases to the baseball case. For, as mentioned above, the case of a fox hunt does not seem to be similar to the case of a baseball game in this sense. Nor does the case of harpooning a whale seem to be similar to the baseball case in this sense. Trying to catch something is a similarity, but this is only one element that ties these cases together as precedents. If you look at the overall pattern recognition type of similarity of the baseball case and the harpooning case, they are not visibly similar at all. They are similar only in some respects. This observation suggests we look at another version of the scheme for argument from analogy.

Guarini has presented a scheme for argument from analogy that he calls the core scheme (Guarini, 2004, 161); a and b are individual objects.

Premise 1: a has features $f_1, f_2, \ldots, f_n$.

Premise 2: b has features $f_1, f_2, \ldots, f_n$.

Conclusion: a and b should be treated or classified in the same way with respect to $f_1, f_2, \ldots, f_n$.

The core scheme fits the arguments from analogy between the animal cases and the baseball case on the basis that the two premises imply that the cases at issue are similar in certain significant respects. A good feature of the core scheme is that it allows the overall dissimilarities between pairs of cases to be overlooked, if the two cases are similar in one or two relevant respects, such as catching something and possessing it. The assumption that the two cases are similar is only implicit, however. It is not stated as a premise in the scheme, and is not necessarily a part of it.

A more specialized scheme for argument from analogy called version 2 in Walton, Reed and Macagno (2008, 58) is built on the simple version, and does have an explicit statement of similarity as its first premise.

Similarity Premise: Generally, case C_1 is similar to case C_2.

Base Premise: A is true (false) in case C_1.

Relevant Similarity Premise: The similarity between C_1 and C_2 observed so far is relevant to the further similarity that is in question.

Conclusion: A is true (false) in case C_2.

The problem with this version of the scheme for argument from analogy is that it does not appear to be a good fit for the arguments from analogy of the kind illustrated in the examples. It depends not only on one similarity premise, but also on another one that may not be easy to apply to cases such as the baseball case and the animal cases. Two questions about the relevant similarity premise need to be answered before this version of the scheme can be applied to the similarities thought to hold between the baseball case and the animal cases. First, what does 'relevant' mean here? Guarini (2004, 162) tells us that he did not include the term 'relevance' in the core scheme because it is common practice not to include relevance claims in argument reconstruction. Second, what is the further similarity? This latter expression suggests that the existing similarity can be reused in future cases. To explore this idea, we turn to case-based reasoning, a technique that reuses a past case to draw conclusions from a current case that is similar in certain respects.

The methods for employing argument from analogy in case-based reasoning in computing uses aspects in which two cases are similar or different are called dimensions and factors. The HYPO system (Ashley, 1988) determines how similar a current case is to past cases by having the relevant similarities each form a dimension. A dimension is a relevant aspect of the case that can take a range of values that move along the scale with values that support one party at one end and the other party at the other end of the scale. In the animal cases, possession, ownership and motive would be examples of dimensions. These dimensions can range on a scale. For example, a dimension might range through cases where the animals are roaming free, cases where the chase had just been started, cases where pursuit was under way, cases of mortal wounding and, finally, at the other end of the scale, to bodily possession. Once determined in a given case, a dimension will favor either the plaintiff or the defendant in a legal case to some degree. For example, in the fox case the plaintiff was in hot pursuit. In the ducks case, the plaintiff was acting for economic gain, while the defendant acted from malice. In the baseball case, both parties were motivated by money, and the plaintiff would have most likely secured the ball had it not been for the assault of the crowd. Bench-Capon (2009, 46) has presented a list of four such dimensions in the wild animal cases and the case of *Popov v. Hayashi*, and ranked them from most pro-plaintiff to most pro-defendant.

CATO is a simpler case-based reasoning system (Aleven, 1997) that was originally designed to aid the teaching of law students. It is based on factors, which can be seen as points on a dimension. In the wild animal cases the following would be factors: whether the party had caught the animal or not, whether the party owned the land or area where the animal was, whether the party was engaged in earning his or her living, and whether the two parties were in competition with each other. Factors are evaluated as arguments favoring one side or the other in relation to social purposes. For example, if the party was engaged in earning his or her living, that would advance the social purpose of the protection of valuable activity. Or if the two parties were in competition, that would advance the social purpose of promoting free enterprise.

Guarini's version of the core scheme for argument from analogy has, instead of a general premise, a premise that states that the two cases being compared share features that should be treated or classified in the same way. These features can be identified with dimensions or factors, depending on whether you are using a HYPO-style system or a CATO-style system. If features that should be classified in the same way are equivalent to respects in which two cases are similar, a simpler version of the core scheme can be cast into the following format, which could be called the single respect scheme.

Respects Premise: Case C_1 is similar to case C_2 in a certain respect.

Base Premise: A is true (false) in case C_1.

Conclusion: Support is offered to the claim that A is true (false) in case C_2.

Where a number of respects are listed, this version becomes equivalent to the core scheme. The conclusion makes it clear that this is a defeasible form of argument in which further evidence can be introduced that can go against or even defeat the argument. This can happen in case-based reasoning, for example, when some factors support A, but then other factors are introduced that support not-A. Then to weigh the arguments on each side, we have to consider the cases on each side, and determine which cases are more on-point, or relevant, that is, the extent to which a case's set of factors covers or overlaps the set of factors in the case at issue. These remarks suggest that to make the core scheme useful for case-based reasoning, we need to bring in a dialectical framework where there is opposition between two opposed claims, of the kind typical in a legal trial, for example.

Typically, in this kind of format, we have an argument from analogy or precedent that supports claim A made by one side, and then on the other side an opposed argument from analogy or precedent that supports claim not-*A*. To comparatively weigh the strength of the one argument as compared with the strength of the opposed argument, we have to bring in something like dimensions or factors that identify the respects in which one case is similar to the other and to have some device for estimating how similar one is to the other by attaching weights to similarity.

In fact, case-based reasoning is built on a kind of method that is dialectical in nature. For example, HYPO is a case-based system that uses dimensions in a format called three-ply argumentation (Ashley, 1988, 206). In the first step, an argument for one side is put forward that matches the past case with the desired outcome and that also matches the case at issue. In the absence of a response, this argument implies that the side putting forward this move should win the dispute. The justification takes the form of an analogy. At the second step, the other side can reply by finding a counterexample, a case that shares the same set of dimensions with the case at issue as the cited case but has opposite outcome, or by distinguishing a case. Distinguishing a case means citing dimensions present in the case at issue that are absent in the case it is compared with and that favor the opposite conclusion, and dimensions in the compared case that favor its outcome that are not present in the case at issue. This move is a rebuttal to the argument of the first move. In the third step, the first party has an opportunity to rebut the distinction, offering a rebuttal to a rebuttal by finding other examples that suggest a different conclusion or by citing cases that defend his or her position.

Wyner and Bench-Capon (2007) devised a system of case-based reasoning that includes a set of six argument structures they describe as argumentation schemes. For example (143) their main scheme (AS_1), looks like the following, where *P* is the plaintiff, *D* the defendant, P_i are the factors, *CC* is the current case and *PC* is the precedent case.

P Factors Premise: P_1 are reasons for *P*.

D Factors Premise: P_2 are reasons for *D*.

Factors Preference Premise: P_1 was preferred to P_2 in PC_i.

CC Weaker Exception: The priority in PC_i does not decide *CC*.

Conclusion: Decide *CC* for *P*.

The factors are rated on a preference scale, and these preferences are used to derive the conclusion. It may be, however, that this scheme does not represent argument from analogy. This scheme, as well as the other five schemes Wyner and Bench-Capon employ in their system, look more like factor-based species of argument from precedent. This brings us to the scheme for argument from precedent and its relation to argument from analogy.

The most common type of argument from precedent used in legal reasoning applies to a current case, and a prior case that has already been decided is taken as a precedent that can be applied to the current case

(Schauer, 1987). The argumentation scheme appropriate for this type of argument is the following one.

Previous Case Premise: C_1 is a previously decided case.

Previous Ruling Premise: In case C_1, rule R was applied and produced finding F.

New Case Premise: C_2 is a new case that has not yet been decided.

Similarity Premise: C_2 is similar to C_1 in relevant respects.

Conclusion: Rule R should be applied to C_2 and produce finding F.

In the baseball case, the rule that was applied is the one called Gray's Rule, the rule that to have possession of the ball, the actor must retain control of it after incidental contact with people and things (McCarthy, 2002, 8). In the baseball case, this rule was applied in a negative way. In the animal cases, the rule was set in place that if you don't catch something, by retaining control of it, you do not fit the requirements for possessing it (in the context of comparing the animal cases and the baseball case). The same rule was then transferred to the baseball case.

Note that this scheme for argument from precedent is built on an underlying argument from analogy represented by the basic scheme (Walton, Reed and Macagno, 2008, 72). This way of configuring the two schemes makes argument from precedent a species of argument from analogy. An advantage of the basic scheme is that it has allowed us to show how some revealing relations among the schemes are involved in case-based reasoning. In the next section, we will see how argument from classification is an extension of argument from analogy typically used in many arguments from precedent.

On this basis, AS_1 can be taken to be a special instance of argument from precedent of the kind specified by the scheme just above. It represents a special subtype of argument from precedent that is designed for use in systems of case-based reasoning that employ dimensions for weighing the respects in which two cases are similar. The core scheme for argument from analogy seems to better represent case-based reasoning techniques using factors or dimensions than the basic scheme, since the core scheme specifically represents respects in which two cases are similar.

3. Arguments from Classification and Definition

Guarini (2004, 162) argues that the core scheme does not fit all cases of argument from analogy. He postulates a second scheme for argument from analogy by extending the core scheme to the next one, which we will call the derived scheme (162):

Premise 1: a has features $f_1, f_2, \ldots, f_n$.

Premise 2: b has features $f_1, f_2, \ldots, f_n$.

Premise 3: a is X in virtue of $f_1, f_2, \ldots, f_n$.

Premise 4: a and b should be treated or classified in the same way with respect to $f_1, f_2, \ldots, f_n$.

Conclusion: b is X.

The derived scheme is in effect a chain argument that is constructed by incorporating the conclusion of the core scheme as an additional premise (premise 4) and adding a new premise (premise 3). The conclusion then says that individual b fits under the category (predicate) of being an X. Thus a way to reconfigure the derived scheme is as follows.

Premise 1: a has features $f_1, f_2, \ldots, f_n$.

Premise 2: b has features $f_1, f_2, \ldots, f_n$.

Conclusion 1: a and b should be treated or classified in the same way with respect to $f_1, f_2, \ldots, f_n$.

Premise 3: a is X in virtue of $f_1, f_2, \ldots, f_n$.

Conclusion 2: b is X.

The first three steps represent the core scheme for argument from analogy, and all five steps, taken together as a chain of reasoning, represent the derived version. This way of proceeding enables us to represent the classification of some individual entity under a general category, which is a feature of some arguments from analogy. Classification is very important as part of the argument, but it needs further amplification to show how classification is tied to argument from analogy in the baseball case.

Recalling the details of Judge McCarthy's analysis of the reasoning in the baseball case, he said that the task of the court was taken to be to craft a definition of 'possession' that applies to the circumstances of the case (McCarthy, 2002, 6). This remark sets in place the first criterion for similarity between the baseball case and the precedent animal cases. All are about the fundamental issue of possession. The problem was that the distinguished law professors disagreed on how possession should be defined. Judge McCarthy then pointed out that although the term 'possession' appears throughout the law, its definition varies depending on the context in which it is used. This situation is not unique to the baseball and animal cases. It is typical of legal reasoning of the kind used in trials, as suggested by Hart's famous example of deciding whether a skateboard is a vehicle that ought to be banned from the park (Hart, 1949; 1961; Loui, 1995). To someone not familiar with disputed cases in legal reasoning, the problem looks easy to solve. It looks as if all we have to do is to define the concept of vehicle. But in hard cases, it is not possible to give a legal definition that provides sufficient support by itself to arrive at a decision that resolves the dispute. The underlying reason is that legal concepts such as 'vehicle' are open-textured, to employ Hart's term, or defeasible, to employ the term currently in used in logic and computing.

As Judge McCarthy put it, the task of the court was taken to craft a definition of 'possession' that applies to the circumstances of the case. But how can this be done given the conflicting opinions on how 'possession' should be defined in law? Law articulates rules or principles that (1) are sometimes established by the courts based on previous cases and that (2) in other instances may even be based on commonly accepted practices that have found their way into law in supporting the formulation of such rules. A set of such rules can provide necessary or sufficient conditions that function as partial definitions. These rules help the argumentation to move forward even in the absence of a fixed definition that is complete and that can be mechanically applied to hard cases. The reader will recall from the description of Judge McCarthy's reasoning above that he used Gray's Rule, the rule that to have possession of the ball, the actor must retain control of it after incidental contact with people and things. Gray's Rule was in turn based on a set of rules for the first possession of baseballs, based on customs and accepted practices in baseball. As applied to the baseball case, this rule led to the conclusion that Popov did not have possession of the ball. However, in the end, even that finding did not resolve the issue of which party had rights to ownership of the ball.

To analyze how the arguments from precedent from the animal cases to the baseball case are based on a notion of similarity that fits the similarity premise of the scheme for argument from precedent, we need to examine some other argumentation schemes that are also involved in the baseball case. The first is the scheme for argument from verbal classification (Walton, Reed and Macagno, 2008, 319).

Individual Premise: *a* has property *F.*

Classification Premise: For all *x*, if *x* has property *F*, then *x* can be classified as having property *G.*

Conclusion: *a* has property *G.*

The case of the drug-sniffing dog (Brewer, 1996) shows how an argument that has been classified in the law literature as argument from analogy is really an instance of arguing from analogy to a verbal classification. Suppose that a trained dog sniffs luggage left in a public place and signals to the police that it contains drugs. Should this event be classified as a search according to the Fourth Amendment? If it can be classified as a search, information obtained as a result of the dog sniffing the luggage is not admissible as evidence. If it is not classified as a search, the information is admissible (Weinreb, 2005).

Ashley's method of distinguishing between deep and shallow analogies between pairs of cases uses an ontology (Ashley, 2009, 8) to represent classifications of concepts to support legal reasoning about claims and issues. This ontological framework specifies and organizes classes of concepts that can be used to represent the important features of cases. It includes representation of actual concepts such as 'animal', as well as legal concepts such as 'possession'. I take this as evidence to support the view that arguments from analogy, as used in law, are based on argument from classification, even though the use of argument from classification in the sequence of reasoning may not be all that obvious in many instances.

On Brewer's analysis, this first classificatory stage of reasoning by analogy leads to a later evaluation stage in which the given event is compared with other cases that have already been classified legally as being searches or as not being searches. Ideally, we could define the term 'search' by using a set of necessary and sufficient conditions for what constitutes a search in any given case, and then apply the definition to the case at issue. Then we could use the argumentation scheme for argument from definition to verbal classification (Walton, Reed and Macagno, 2008, 319).

Definition Premise: a fits definition D.

Classification Premise: For all x, if a fits definition D, then x can be classified as having property G.

Conclusion: a has property G.

However, although this scheme may work in easy cases, where the definition clearly fits the case, it falls down when the term at issue is defeasible. Then what we need is a defeasible definition, but since the definition is defeasible, it may still be open to contention what conclusion it directs us to draw in the case at issue. As the baseball case shows very well, argument from definition to verbal classification does not work, and we have to fall back on Gray's Rule. Since defeasible definitions are ubiquitous in legal argumentation, as Hart showed (1949; 1961), and as the cases treated here illustrate, these considerations bring out the importance of modeling them in some way that is both precise and useful.

The theory of defeasible definitions provided by McCarty and Sridharan (1982) uses what are called prototypes and deformations. On this approach, there is an invariant component to provide necessary, but not sufficient, conditions for the existence of the concept, a set of exemplars, each of which matches some but not all of the instances of the concept, and a set of transformations in the definitional expansion that expresses relationships between the exemplars. McCarty and Sridharan state that one exemplar can be mapped into another exemplar in a certain way. This method of working with defeasible definitions in argumentation in artificial intelligence and law has been applied to the case of *Eisner v. Macomber* (252 U.S. 189 (1920)).

4. Similarity

The basis for deciding whether one case is a precedent for another in law has been the subject of debate for generations, and a common view is that a precedent case holds for cases that are similar but not identical to it (Schauer, 2009, 46). How this works is easy to see if two cases are very similar in obvious respects, but how is a case where a man sued a company because there was a decomposed snail in his beer bottle similar to a case where a man tried to sue because of a defective Buick automobile? The answer is that even though the two cases are dissimilar in many respects, they are similar in that they were both consumer transactions that caused harm, and the defect was not immediately apparent (Schauer, 2009, 46). But surely, just these common respects are not enough in themselves to make the one case similar enough to the other so that one could be taken as a precedent for the other. There is something about the common sequence of events that makes the one case similar to the other. First the plaintiff bought some product that he assumed was the normal product he expected, and he thought therefore that the product was reasonably safe to use. Then something in the product turned out to be defective, and when he used the product this defect caused some harm that impacted badly on his health. There is a thread, or sequence of events, that is of the same kind in both cases. It started in the same way, went through the same kind of chain of events and ended in the same way. Another thing both cases have in common is that both were about recompense that the plaintiff claimed was due to him because of harm he supposedly suffered. They are both about the same basic issue that defines the claim to be proved in the lawsuit.

What is the similarity between the wild animal cases and the baseball case that enables an argument from precedent to be drawn from the one to the other? The situation of a baseball hit into the stands where fans jostle to try to retrieve it is not similar to a situation of fishing for a whale or hunting a fox. Gray (2002, 1), as quoted in Section 1 of this chapter, made the point that catching a baseball is not similar to mortally wounding a fox: "a baseball at the end of its arc of descent is not at all like a fox racing across the commons, acting under its own volition, desperately attempting to evade death at the hands of its pursuers". Even though the animals are different, and the details of how they are caught or pursued are different, the wild animal cases are similar among themselves. They are all about pursuing, catching, wounding and holding wild animals, and about which party has the right to possession of the animal at the end of the process. In most of these respects, the whale cases, the fox case and the ducks case are similar. They are all about this same process of pursuing and possessing wild animals. The baseball case is noticeably different. It is not about pursuing, catching or possessing a wild animal.

So what similarity is there that supports the transfer via the arguments from precedent from the earlier ones to the later one? It is not just the element of possession, for there are many cases of disputes about possession of something that are not similar enough to these cases to provide precedents for them. The similarity involves both possession and this pursuing and catching process. All the cases are about catching something, or attempting to catch it, and about which party may rightly be said to possess it at the end of this attempting-to-catch process. They are also about someone else interfering with this process and preventing the other party from catching and possessing the animal.

When you abstract from the details of the animal cases and the baseball case that are not relevant in the argument from analogy that connects them, what is left is a template linking a series of events and questions into an ordered sequence. If we distinguish following Ashley (2009) between deep and shallow analogies, a template that matches up the same sequence fitting two cases can reveal a deep similarity that is more significant, as opposed to a shallow similarity in which the two cases do not appear to be similar. The sequence template for the deep similarity that runs through the animal cases and the baseball case is visually represented in Figure 5.1. In the next section, it will be shown how such sequence templates can be represented as abstract structures that can be applied to real cases of arguments from analogy.

If this analysis of this special type of similarity between these pairs of cases is correct, the consequences for studying how argument from precedent is based on argument from analogy are highly significant. When we say that two disputed cases are similar, and therefore that the one case can work as a precedent for the other, it does not mean that the two cases appear to be similar in many respects, so that there has to be a visual match of some sort between them. This pattern recognition kind of similarity represents only a superficial type of similarity. Superficially, the cases initially look very different. It looks like there is no basis for a compelling argument from analogy between them. It is only when you probe into them further, detect a sequence in how the concepts in each case are tied together in a template within the argumentation about the dispute at issue and see how this template affects the reasoning on each side that the similarity important for precedent emerges. McLaren (2003; 2006) has developed a two-stage case retrieval system in SIROCCO that assessed similarity of cases in terms of sequence of events and demonstrated empirically the utility of the approach in improving retrieval of relevant engineering ethics cases involving engineering ethics code provisions. The template is just one small part in the larger structure of the dialogue in a case (Ashley, 2004; 2009, 9) that goes through several stages.

These observations suggest that there are three stages to using argument from analogy. At the first stage, two cases may look similar, and this apparent match may suggest a rough analogy that could be used to support an argument from analogy. At the second stage, a closer look at the similarity premise can be given, to see whether the similarity is merely visually apparent, as an instance of pattern recognition, or whether there is a logical similarity of the kind that can be supported by applying a template like that pictured in Figure 5.1. The third stage is the evaluation of the

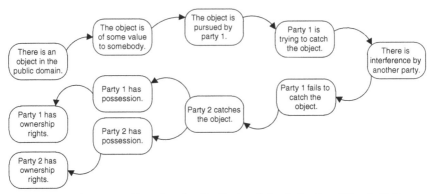

FIGURE 5.1 Sequence Template for Similarity of the Animal and Baseball Cases

argument from analogy, by citing and comparing the respects in which the one case is similar to (or dissimilar from) the other. Although the basic scheme for argument from analogy is initially the more useful for identifying this type of argument at an early stage, when you get to the later stages of analyzing and evaluating arguments from analogy, the core scheme becomes more useful.

In these cases, you have to recognize that the dispute is about possession of the contested entity, and the reasoning relates to details of this attempting-to-catch process, how it went along the way as the two parties took part in it and how it ended up. The similarity between these cases that supports argument from analogy and argument from precedent needs to be represented by a typical sequence of actions, events and questions of the kind shown in Figure 5.1. Seven steps in the sequence are episodes, and the last two are questions. In this example, the first seven steps represent a sequence of intentional actions of an agent, and something that interferes with the agent's achieving his goal.

5. Scripts and Stories

Commonly known ways of carrying out everyday activities were codified in early work in artificial intelligence (Schank and Abelson, 1977) in sequences called scripts. The standard example is the restaurant script, an ordered set of seven statements: (1) John went to a restaurant. (2) The hostess seated John. (3) The waitress gave John a menu. (4) John ordered a lobster. (5) He was served. (6) He left a tip. (7) He left the restaurant. Gaps in the sequence can be made explicit by defeasible inferences based on common knowledge about the way things are normally done in the script. For example, we can infer defeasibly that lobster was on the menu. It would be an exception if lobster was a special item not listed on the menu, and the waitress told John about it. However, the gap-filling inference can be drawn if there is no information to the contrary, because restaurant customers normally get their information about what to order from the menu.

Modules called memory organization packages (MOPs) (Schank, 1986), which also represent stereotyped sequences of events, are used in case-based reasoning (Leake, 1992). They are smaller than scripts and can be combined in a way that is appropriate for the situation in which they are needed. For example, the space launch MOP includes a launch, a space walk and a re-entry (Leake, 1992, 73). Scripts and MOPs can be used to build or amplify what is often called a story, a connected sequence of events or actions that hangs together, that is ordered as a sequence and that contains gaps that can be filled in.

Pennington and Hastie (1993), among other authors, have argued that understanding actions carried out in criminal cases is done by constructing competing stories about what supposedly happened using the evidence in the case. The method is to find the best story, the best script connecting the known facts, or at any rate the one that seems most plausible based on the evidence. Such a plausible story describes a general pattern of states or kinds with which we are all normally familiar. The problem is that a plausible story may not be very well supported by the evidence, whereas a less plausible story may be supported by more evidence. To deal with this problem, Wagenaar, van Koppen and Crombag (1993) devised a special type of story used to represent legal reasoning called an anchored narrative. Bex (2009b) has proposed a hybrid framework for reasoning with arguments, stories and criminal evidence, a formal framework that shows how the plausibility of the story can be evaluated by giving arguments that ground the story on evidence that supports or attacks it.

Pennington and Hastie (1993) also had the idea that the plausibility of a story can be tested by its evidential support. They devised the notion of an episode scheme, which is like a script or MOP except that it can be more abstract or more specific. An example would be a scheme for intentional action that describes the general pattern of events in the restaurant script, by citing the events of ordering, eating and paying (Bex, 2009a, 94). Bex (2009b) combined the episode schemes of Pennington and Hastie with the scripts of Schank and his colleagues to form what are called story schemes. These are modeled as an ordered list of events or types of events that can be more abstract or more specific. Bex (2009b, 59) offers the following example. John Haaknat is a drug addict who needs money and decides to rob a supermarket. He gets the money and jumps into his car and takes off, but seeing the police he parks his car at a nearby park and then jumps into a moat to hide. Later the police search the park and find him soaking wet from the water in the moat. Bex (2009b, 59) constructs a graph that exhibits the causal relations between the various events in the story, as shown in Figure 5.2.

Bex (2009b, 59) calls it a causal structure, because it contains implicit causal relations assumed by the reader of the story that enable the reader

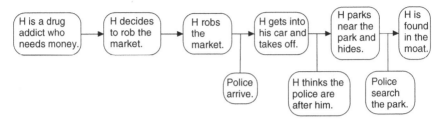

FIGURE 5.2 Causal Structure of the Haaknat Story

to connect the sequence as a series of events and actions that make sense. We can recognize it as a story, even though not all the events and causal relations have been rendered explicitly.

6. Modeling the Sequence Template as a Story Scheme

Evidential reasoning in law is typically based on general knowledge accepted in a certain community, codified in the form of generalizations (Bex, Prakken, Reed and Walton, 2003). Examples of such generalizations are 'the forceful impact of a hammer can cause a person's skull to break' and 'witnesses under oath usually speak the truth' (Bex, 2009b, 18). Generalizations can default when applied to specific instances. For example, it may not be true that the forceful impact of a plastic hammer can cause a person's skull to break. A story scheme is a collection of literal schemes and (causal) generalizations schemes that fits the following definition (Bex, 2009b, 126).

> **Definition**: A story scheme is a set comprised of literal schemes and causal generalizations such that the set of components ( ) = { } or is the antecedent or the consequent of some.

Both generalizations and story schemes can be abstract as well as specific. The underlying logic of this framework is based on a set of inference rules for classical logic combined with a defeasible *modus ponens* rule for the conditional operator $\Rightarrow$ that represents defeasible generalizations (Bex and Prakken, 2010). A generalization has the form p_1 & p_2 & ... & $p_n \Rightarrow q$. A generalization with free variables is a scheme for all its ground instances, and a literal scheme is a scheme for all its ground instances. For example (Bex, 2009b, 126), 'x robs y' is a scheme for 'Haaknat robs supermarket' and also a scheme for 'John robs bank'. A story scheme can also contain causal links, as in the following example: {motive $\Rightarrow$ C goal, goal $\Rightarrow$ C action, action $\Rightarrow$ C consequence}.

A set of events or actions in a story corresponds to a component of the story scheme if the scheme is derivable from the events through a process of applying abstractions. This process of linking to particular events or

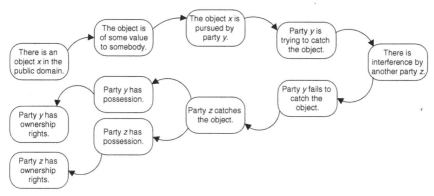

FIGURE 5.3 Story Scheme That the Baseball Case Shares with Animal Cases

actions described in the story to their representation in a more abstract level by a story scheme is explained by Bex (2009b, 127) with two examples. In the first example, the event 'Haaknat robs supermarket' is a particular instance of the abstract scheme '*x* robs *y*' straightforwardly, and a more complex inferential process that Bex calls an explicit abstraction generalization is not needed. In the second example, however, a more complex process is required. In this example, 'Haaknat robs supermarket' is said to correspond to the action component of the intentional action scheme through the abstraction generalization Haaknat robs supermarket ⇒ *A* action because {Haaknat robs supermarket} {Haaknat robs supermarket ⇒ *A* action} ⊢ *A* action.

The sequence template shown in Figure 5.1, classified as an episode scheme in the sense of Pennington and Hastie (1993), can also be seen at a higher level abstraction as a story scheme in the sense of Bex (2009a; 2009b), as we now show.

The story scheme shown in Figure 5.3 is an abstraction, a template that offers a way of representing the sequence of actions in the template at a higher level of abstraction representing a story as a connected causal sequence.

In the cases discussed in this chapter, the story scheme in Figure 5.3 links the fox case, the fish case and the whale case all together as similar in relevant respects, and links each of them to the baseball case. This does not mean that each case is similar to each other in every respect, or even in every relevant respect. It means only that they are similar in that they all share a certain abstract pattern as a story scheme that fits the general causal sequence represented in Figure 5.3.

At this point we have represented the notion of similarity that argument from analogy is based on, by adapting the story scheme structure of Bex (2009b) that evolved from the story-based approach to reasoning about factual issues in criminal cases. This analysis offers a better way of showing how to marshal evidence in support of the similarity premise in an argument from analogy. It also offers support for the view that version 2 of the argumentation scheme for argument from analogy applies better to cases such as those exploiting the analogy between the animal cases and the baseball case than version 1. However, we still have not posed the question of how to evaluate arguments from analogy that fit version 2 of the argumentation scheme. Some clues as to how to go about this are suggested by the dialogue structures used by Bex (2009b, 139–156) and by the formal dialogue game designed for use in investigating stories by Bex and Prakken (2010). Arguments from analogy are defeasible, because even though an analogy can be strong in certain respects, it can always be attacked by showing that it fails to hold, or is weak, in other respects. It is suggested in the next section that some directions on how to approach the problem of evaluation of arguments from analogy are offered by finding some resources from case-based reasoning and from these dialogue models.

7. Dialectical Aspects of Argument from Analogy

Using the Haaknat example again, Bex and Prakken (2010, 5) show how two competing explanations that are offered as evidence in a criminal case can be evaluated to see which is the better explanation. One criterion they use is evidential coverage, meaning how many arguments can be used to support claims that are parts of the explanation. Haaknat was found hiding in a moat in the park after the robbery, and the prosecution explanation was that he had fled there after the robbery to avoid arrest. Haaknat offered a different explanation. He argued that he was hiding in the moat because he had an argument with a man over some money, and this man had drawn a knife. Haaknat's explanation was that he had fled to escape this man. There are various criteria that can be used to evaluate which is the more plausible explanation, internal consistency of each story being one of them. Bex and Prakken provide a formal dialogue model that represents a process of evaluation in which each side presents arguments to support its own story and asks critical questions to test and throw doubt on the plausibility of the other party's story. The same kind of dialogue model can also be used to provide a method for evaluating the strength of an argument from analogy, the case in point being the analogy between the baseball case and the previous animal cases.

Judge McCarthy (2002, 9) ruled that although Popov did not retain control of the ball, other factors need to be considered. One factor is that his efforts to retain control were interrupted by a violent mob of wrongdoers. Another is the principle that if an agent takes steps to achieve possession of a piece of abandoned property, but is interrupted by the actions of others, he has a pre-possessory interest in the property. After examining all the arguments, Judge McCarthy decided that any award to one party would be unfair to the other and that each had an equal and undivided interest in the ball. In the end, the precedents from the animal cases did not decide the outcome of the case. But still, they did help to support Gray's Rule, and Gray's Rule acted as a partial definition of possession that influenced the line of reasoning that led to the decision. So argument from precedent and argument from analogy, as well as argument from classification and definition, were important in understanding how the sequence of argumentation in the case went.

The baseball case suggests that argument from analogy cannot be analyzed and evaluated in specific cases without placing its use within a broader context where there is a disputed issue. This context includes a sequence of argumentation relevant to that issue that it intended to resolve by weighing the arguments on both sides. How does the process of applying these schemes to evaluating the arguments by fitting them to cases in this context work?

- The process uses general rules derived from legally authoritative sources by statutory interpretation.
- It uses arguments from analogy to previous decided cases.
- Argument from precedent is based on argument from analogy.
- It uses argument from established rules from these sources.
- In some instances, it uses argument from generally accepted practices in specific kinds of practical activity domains.
- It uses and arrives at classifications based on these rules.
- Instead of fixed definitions, it uses defeasible partial definitions in the form of necessary and sufficient condition rules.
- It applies these rules to the problematic case that needs to be decided by examining and weighing the arguments pro and contra based on the evidence from these and other sources.

The argumentation in a trial can be viewed in this context as a pro-contra dialogue process in which one side puts forward arguments, the other side puts forward opposed arguments, and then each side gets a chance to critically examine the claims and arguments of the other side. Critical questioning, therefore, as well as argumentation schemes representing the different types of arguments, are both important. The task of weighing the arguments requires looking at how each argument can be questioned and attacked (Atkinson et al., 2004). Matching each scheme there is a set of typical critical questions that can be used to reply to an argument of that type by probing into its weak points.

The dialogue game of Bex and Prakken (2010) is designed to regulate the discussion in a criminal case where both players want to find a plausible explanation for the facts of the case, and where the goal is to find the best explanation. Each competing explanation is modeled as a story that can be supported or attacked by the factual evidence in the case, and also evaluated by other criteria such as internal consistency. Even though in this chapter the central concern is not explanation but argument from analogy, this dialogue game is useful because it contains arguments and easily accommodates the use of defeasible argumentation schemes (Prakken, 2005). In such a dialogue, or another comparable type of dialogue of the kinds used in artificial intelligence and law, when an argument is put forward it can be attacked in several ways. When an argument from analogy is initially put forward, it is possible that there is a strong or even striking similarity between the case at issue and the analogous case. As the dialogue proceeds, however, questions may arise as to whether the two cases are similar in certain specific respects or dissimilar in other respects. It is a sequence of argument moves during a particular stage of a dialogue that determines how strong the argument from analogy should be taken to be, from a logical point of view. It is this dialogue sequence that should provide the basis for evaluating the strength of the argument from analogy.

As noted above, HYPO processes cases based on arguments from analogy and precedent using the process called three-ply argumentation (Ashley, 1988, 206). First, an argument for one side is constructed by finding a past case in which the outcome closely matches that of the desired outcome of the case under consideration, based on the dimensions. Second, the other side can reply in one of several ways. The other side can reply by finding a counterexample, a past case that matches the current case but that has the opposite outcome. Another reply is to distinguish the case by pointing to dimensions present in the current case that are absent in the precedent. Third, the original party can offer a rebuttal of the previous move by making several kinds of moves. These include distinguishing counterexamples, pointing out additional dimensions or citing cases that show that weakness identified does not rebut his position. The three-ply argumentation could be used to effectively set up the pre and post conditions for a dialogue model of HYPO, for example, by specifying sets of critical questions for argument from precedent.

The three-ply HYPO sequence can be compared to the set of critical questions matching version 1 of the argument from analogy (Walton, Reed and Macagno, 2008, 315).

CQ_1: Are there differences between C_1 and C_2 that would tend to undermine the force of the similarity cited?

CQ_2: Is A true (false) in C_1?

CQ_3: Is there some other case C_3 that is also similar to C_1, but in which A is false (true)?

CQ_1 corresponds to the reply in HYPO of distinguishing the case by pointing to dimensions present in the current case that are absent in the precedent and that favor the opposite conclusion, and dimensions in the precedent that favor its outcome that are not present in the current case.

CQ_3 corresponds to the reply in HYPO of finding a counterexample. The reply of offering a rebuttal in HYPO fits under CQ_2. The reply of citing cases that show that weakness identified does not rebut his position is not illustrated in the baseball case, but it could suggest a continuation of the argument by further pro and contra argumentation.

8. Two Other Test Cases and Their Implications

There are many different types of arguments from analogy, and the tools for analysis applied to the Popov case in this chapter fit some better than others. However, in this section it is most useful to deal briefly with two cases. The first one is interesting because its basic structure appears to be fairly simple in the way it fits the story scheme. The second one brings up some important points concerning the relationship between the two schemes for argument from analogy set out in Section 1.

One of the most famous cases of argument from analogy in public affairs is the hypothetical violinist case (quoted below), used to argue that abortion is permissible (Thomson, 1971, 48–49).

You wake up in the morning and find yourself back to back in bed with an unconscious violinist, a famous unconscious violinist. He has been found to have a fatal kidney ailment, and the Society of Music Lovers has canvased all the available medical records and found that you alone have the right blood type to help. They have therefore kidnapped you, and last night the violinist's circulatory system was plugged into yours, so that your kidneys can be used to extract poisons from his blood as well as your own. The director of the hospital now tells you, "Look, we're sorry the Society of Music Lovers did this to you – we would never have permitted it if we had known. But still, they did it, and the violinist now is plugged into you. To unplug you would be to kill him. But never mind, it's only for nine months. By then he will have recovered from his ailment, and can safely be unplugged from you."

This hypothetical case has been taken to be highly persuasive as an ethical argument on the abortion issue, even though there is a large amount of literature containing arguments supporting and attacking it. When this case is presented to a respondent, he or she is likely to agree that the person attached to the violinist has the right to unplug himself. According to the argument from analogy, this statement fits together with the similarity premise to enable the conclusion to be drawn that a pregnant woman has the right to terminate her pregnancy, even though the fetus will die as a result. Thomson's argument is that the person in the source case who unplugs the violinist does not violate his right to life, because the violinist has no right to the use of that other person's body. By argument from analogy, we are then led to the conclusion that abortion does not violate the fetus's right to life but merely deprives the fetus of the use of the pregnant woman's body, something to which the fetus has no right.

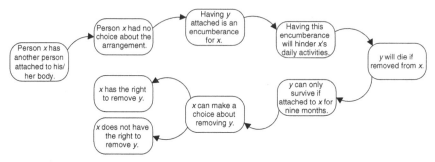

FIGURE 5.4　Story Scheme for the Violinist Case

This case can be analyzed using the following episode scheme: {person x has had another person y attached to his body without x having any choice; having y attached is an encumbrance that will hinder x's daily activities; x and y are attached in such a way that y will die if removed from x; y can only survive when removed from x after a period of 9 months; x can make a choice about whether to have y removed or not}. This episode scheme is shown in Figure 5.4.

This story scheme presents an abstract structure that applies both to the violinist case and to a case at issue about a woman who has become pregnant. The conclusion drawn from the description of the person attached to the violinist is designed to elicit the idea that the person to whom the violinist was attached should have the right to choose to have him detached. By argument from analogy, the conclusion drawn is that a woman who has become pregnant due to rape should have the right to have an abortion.

The violinist case has been much discussed in the ethical literature, and many arguments pro and contra have been put forward by citing the respects in which the two cases are similar or different. Objections to the argument have tended to proceed by arguing that there are important differences between the violinist case and cases of a mother aborting a fetus. One such objection is that the argument extends only to cases of abortion where the pregnancy was caused by rape. In the violinist case, the person kidnapped did nothing himself to cause the violinist to be attached to him, whereas in typical abortion cases, the pregnant woman chose to have intercourse. Another difference is that the fetus is the woman's child, while the violinist is a stranger.

The argument from analogy initially appears plausible, for two reasons. One is that there appears to be a striking similarity between the two cases because the story scheme ties together a set of common elements in a sequence that both cases exhibit. The other reason is that in the violinist case it seems reasonable to conclude that the person attached to the violinist should have the right to remove him, assuming that he is not obliged in any way to support the violinist by undergoing the arduous procedure that is necessary. When you combine these two reasons, the violinist case appears to present a strong argument from analogy that is in favor of the conclusion it was put forward to support. But as differences are explored as well as similarities, the argument begins to seem less compelling.

The basic problem we started out with in this chapter is that there seemed to be two different argumentation schemes for argument from analogy, and it seems difficult to choose which one is the better or which one should generally be used. One scheme is very simple. It simply states that when two cases are similar, and where some conclusion can be drawn in the one case, a comparable conclusion should also be drawn in the other case. The other scheme is more complex. It says that two cases are similar in certain respects, allowing for the possibility they may also be different in other respects, and then it claims that the respects in which they are similar are decisive, or outweigh the respects in which they are different. On this basis, it claims that because some conclusion can be drawn in one case, a comparable conclusion should also be drawn in the other. The more complex scheme easily fits the case-based models of argument from analogy that use factors or dimensions.

What is the relationship between these two forms of argument? Here the following hypothesis is put forward, based on the approach that when an argument from analogy is first used at some stage of a dialogue to persuade the other party to accept a claim, it needs to be tested and evaluated by means of a sequence of argumentation that follows from this initial move. When the argument is first put forward it is in the structure of the simple version of the scheme, but later during the sequence, as critical questions are posed and counterarguments are put forward, the more complex version of the scheme is the structure that the argumentation best fits.

A test case that is interesting to briefly consider is from copyright law. Striking similarity has sometimes been used as an argument that relies on similarities or claimed identities between two works to prove that there was a violation of copyright law in which one party copied some intellectual property belonging to the other. In law, in order to prove such a copyright violation there are two component claims that have to be proved: copying and improper appropriation of copyrightable expression. In the absence of direct evidence of copying, one may meet the burden of proving copying indirectly by showing that the alleged copier had access to the work and that the two works are substantially similar. The doctrine of striking similarity arises because two works, for example, two songs, can be so strikingly alike in their sounds, their notes, and the sequences of tones and cadences in the melodies that it might seem to someone suspecting his or her work has been stolen that there could be absolutely no doubt that the one has been copied from the other. However, such an appearance of similarity or even identity can be misleading when drawing an inference about copying. There could be other explanations. It could be merely coincidence, or both parties may have copied from a common source in the public domain. Hence striking similarity should not, by itself, be regarded as sufficient evidence to fulfill the burden of proof for establishing violation of copyright. Arguments from analogy based on such a striking similarity nevertheless are highly persuasive, because of the powerful psychological effect of the perceived similarity. The nature of the problem is indicated in Judge Frank's often-cited opinion in *Arnstein v. Porter* (154 F.2d 464 (2nd Cir. 1946)) to the effect that even where access is absent, the similarity can be so striking that it precludes the possibility that the plaintiff and defendant independently arrived at the same result. This ruling could be "disastrous" (Patry, 2005/6) if it were used as a sufficient basis for justifying an inference of copying in all circumstances. The problem is to judge why this kind of direct inference from striking similarity to a conclusion of copyright violation represents a kind of inference that jumps to a conclusion quickly while overlooking other evidence that needs to be taken into account.

Arguments from striking similarity are based on the simple version of the argumentation scheme for argument from analogy. If the analogy between two cases is striking, meaning that it is much stronger and more convincing than the usual kind of comparison between cases, the argument initially appears to be so strong that it is frozen at the initial stage. Instead of moving on to the sequence of argumentation including critical questioning and counterarguments, the sequence stops there. What this shows is that if we go exclusively by the simple version of argument from analogy, the argument may appear so powerfully persuasive that it jumps to a simplistic conclusion without taking into account other relevant evidence that needs to be considered. The hypothesis that both schemes need to be used can help to explain what has gone wrong when an argument from analogy based on striking similarity jumps to a premature conclusion. It initially provides some legitimate evidence to support an ultimate claim based on a highly persuasive analogy, but its evaluation has not proceeded far enough into the subsequent sequence of argumentation in the dialogue to properly take into account all the relevant evidence that needs to be considered. It is precisely for these reasons that jumping to a conclusion is a form of argumentation often associated with logical errors and informal fallacies (Walton and Gordon, 2009).

Carneades is a system for reasoning with argumentation schemes, and has the distinctive feature that it sorts the critical questions matching a scheme into three categories, thereby enabling them to be treated as premises of the scheme, in some instances additional implicit premises (Walton and Gordon, 2005). It would be easy to manage critical questions by modeling them as additional premises in a scheme, except that there are two different variations on what happens when a respondent asks a critical question. In some instances, when a critical question is asked, a burden of proof shifts to the proponent's side to answer it, and if this burden is not fulfilled the argument is defeated. In other instances, merely asking the question does not defeat the proponent's argument until the respondent offers some evidence to back it up. To cope with variation, Carneades distinguishes three types of premises, called ordinary premises, assumptions and exceptions. An assumption is not explicitly stated in the premises of a scheme, but behaves like an ordinary premise, one that was explicitly stated. An assumption is taken to hold, so that if a critical question is directed to it, and some evidence is not given to support it in line with the questioning, it now fails to hold. An exception is not taken to hold unless evidence can be given to show that it does hold. By treating the argumentation schemes and their matching critical questions this way, the Carneades system makes it less crucial whether some factor that is important for evaluating an argument that fits a scheme is treated as a premise of the scheme or as a critical question matching the scheme.

We can classify the critical questions matching version 1 of the scheme for argument analogy, as shown in Section 7, as follows. The second critical question merely asks whether one of the premises is true, so it can be treated as a normal premise. The first and third questions cite specific differences or another case that is needed to furnish evidence required to call the argument from analogy into question. So they are best treated as exceptions.

What these observations reveal is that if we use Carneades to model argumentation from analogy, the critical question that asks whether there are differences between the two cases can be represented as an additional premise of the simple scheme. This appears to show that there is a transition from the simple scheme to the more complex scheme. This transition appears to represent a typical sequence of dialogue in which the argument from analogy is analyzed in greater depth. In the simple scheme, the factors or dimensions appear only in the critical questions, but in the more complex version of the scheme, more of them appear in the scheme itself. From the point of view of the Carneades model, since critical questions can be represented by fitting them into the scheme and treating them as premises, and since counterarguments can be represented in a dialogue format, the structure allows for an orderly transition from the application of the one scheme to the other. But before this transition can be properly understood, both schemes need to be revised.

9. Reconfiguring the Schemes

What is the best argumentation scheme and the best set of critical questions for argument from analogy, from among those surveyed in Section 2? The best one to work with initially is the following modified version of the simplest scheme from Walton, Reed and Macagno (2008, 315).

Similarity Premise: Generally, case C_1 is similar to case C_2, based on their shared story scheme.

Base Premise: A is the conclusion to be drawn in case C_1.

Conclusion: The conclusion B, comparable to A in case C_1, based on their shared story scheme, is to be drawn in case C_2.

The violinist case can be used to illustrate how this argumentation scheme applies to an argument from analogy. The similarity premise is the statement that the case of the person with the violinist attached to his body is similar to the case of a woman with a pregnancy due to rape. The base premise is the statement that the conclusion to be drawn in the violinist case is that the person attached to the violinist should have the right to detach him. The comparable conclusion to be drawn in the pregnant woman case is that she should have the right to an abortion.

In light of the examples studied in the chapter, and especially in light of the use of models in artificial intelligence based on factors and dimensions, the following set of critical questions matching the scheme is now proposed.

CQ₁: Are there respects in which C_1 and C_2 are different that would tend to undermine the force of the similarity cited?

CQ₂: Is A the right conclusion to be drawn in C_1?

CQ₃: Is there some other case C_3 that is also similar to C_1, but in which some conclusion other than A should be drawn?

It should be noted here that the first critical question relates to factors or dimensions that represent similarities or differences between the two cases that tend either to support or to detract from the argument from analogy. Another way to look at this critical question, therefore, is as an initial point in a sequence of dialogue that goes into pro and contra arguments with respect to the claim made in the argument from analogy. The third critical question also represents a kind of counterargument that is often called a counter-analogy in logic textbooks, a second argument from analogy directed against the first one that goes to the opposite conclusion of the first one. The critical questions can be viewed as representing species of counterarguments, and as well, on the Carneades model, they can be viewed as species of premises of the scheme. On this model, their function is to shift the burden of proof from the one side to the other in dialogue. Thus we can see that there are different ways of evaluating arguments from analogy, but the main functions of the simplest version of the argumentation scheme and its matching set of critical questions are to enable us to identify arguments from analogy and to provide at least some entry point for instructing a beginner about questioning them.

Once we have identified and analyzed the argumentation in a given instance as an argument from analogy using the story scheme model of the similarity premise along with the other premise, we typically want to go on to the next tasks of analyzing and evaluating the argument. This takes us to the following sequence of the argumentation following the initial use of the argument from analogy in the dialogue. The best device that is useful at this stage is case-based reasoning with its use of dimensions and factors. The simple scheme above is the best representation of the form of argument from analogy to be used, however, because it distinguishes the respects in which one case is similar to another, based on the story scheme common to the two cases. These respects can then be added to or challenged by bringing out new ones, or new differences between the two cases, that is, factors or dimensions.

During the process of argument evaluation, there is a two-part dialogue sequence representing how an argument from analogy is typically put forward initially and then later critically questioned and examined in more detail for its strengths and weaknesses. In the first part, the argument from analogy appears plausible if there is a story scheme into which the sequence template for the two cases fits. Such a fit makes the argument from analogy appear strong by supporting the similarity premise of the basic scheme. The basic scheme does not distinguish between other additional respects that may be brought out in which the one case is arguably similar to the other or dissimilar. However, during the next part of the sequence, issues concerning specific respects in which the one case is similar or not to the other may arise.

Guarini's version of the scheme for argument from analogy (Section 2) has a premise that states that the two cases being compared share a set of features. By treating features as equivalent to respects in which two cases are similar, a simpler version of the core scheme was recast into what we called the single respect scheme. To fit with the new version of the simplest scheme above, we now offer this reformulated version of the single respect scheme.

Respects Premise: Case C_1 is similar to case C_2 in a certain respect.

Base Premise: A is the conclusion to be drawn in case C_1.

Conclusion: The conclusion comparable to A in C_1 is to be drawn in case C_2.

When a number of respects are brought together by using the single respect scheme, repeatedly citing several common features in which the two cases are held to be similar, then the single respect scheme becomes equivalent to Guarini's core scheme. Respects, or features if you will, can be identified with dimensions or factors, depending on whether you are using a HYPO-style or a CATO-style system.

When the simple scheme is applied to a case, there are already some points of similarity, as well as a general pattern of similarity, postulated by the story scheme. The single respects scheme is supposed to go beyond this level by citing specific features shared by the two cases. To make this scheme useful in relation to the simple scheme, we have to define 'respect' as referring to a specific feature in which two cases are held to be similar. In this way, the single respect scheme is an extension of the simple scheme.

Next, the problem is to fit the simplest scheme and its matching set of critical questions alongside the single respect scheme into a dialogue framework in which they can be employed alongside each other. How this will work is that the new version of the simple scheme is applied at the first point in the dialogue where the argument from analogy is put forward. Next, the critical questions are asked, and the first critical question concerns respects in which C_1 and C_2 may be different. Put in a stronger form, this critical question could be a counterargument that draws a conclusion that is the negation of the original argument from analogy, based on the premise that there are one or more respects in which the two cases are different. This is a contra argument, but there is also a matching pro argument of the type represented by the single respect scheme. So the single respect scheme represents a form of argument that is opposed to the counterargument based on different respects. Such a pattern of argumentation is common in case-based reasoning where some factors support the conclusion A, but then other factors are cited that support the conclusion not-A. As noted, case-based reasoning weighs the arguments on both sides by considering the cases on each side and determining which cases are more on point. To be on point is to be relevant in terms of the overlap of each case's factors with those of the current case. Fitting together these arguments and critical questions needs to be done in a dialectical framework where an argument from analogy is initially put forward and then challenged, first by asking critical questions, but then at a later stage by probing into specific respects in which two cases are similar or different. What we see is that there is a surface level that represents the initial impact of putting forward an argument from analogy, and a deeper level of analysis and criticism in which specific respects of similarity or difference are specified in weight against each other.

Our new version of the simplest scheme for argument from analogy has three components: the similarity premises, the base premise and the conclusion. The single respect scheme and the multiple variants of it corresponding to Guarini's scheme are addressed to the similarity premise. Depending on whether the respect cited is a pro or contra factor, the respect scheme will be an argument for or against the similarity premise. Notice that when an argument fitting such a respect scheme is brought forward, we are now at the evaluation stage. When the argument from analogy is originally brought forward in the form of an argument fitting the simple scheme, we are just at the presentation stage where the argument, if the similarity fits a plausible story scheme, will be a provisionally acceptable argument. But when the dialogue starts to go into a discussion of specific similarities and differences between the two cases, we have now entered the evaluation stage. In other words, during this stage the question being discussed is how plausible the argument is. The important thing to note is that when the specific respects in which the one case is similar to or different from the other are being put forward, the arguments are either supporting the similarity premise or attacking it. The first critical question of the simple scheme represents this kind of attack on the similarity premise, whereas the third critical question represents a different kind of argument. In this different kind of argument, the counterargument purports to prove the negation of the conclusion of the simplest scheme.

10. Conclusions

Judge McCarthy's remark cited at the beginning of Section 7 shows that the fundamental task of the trial in the baseball case was to craft a definition of possession that would be applicable to the case. The court partly carried out this task by using defeasible rules for partly defining possession of baseballs. But these rules failed to resolve the issue of which party should legally have ownership of the contested baseball. What the trial showed is that although the baseball case is about argument from classification and definitions, as well as about argument from analogy, the argument from definition cannot solve the problem by itself. It cannot be solved by itself because the concept of possession that needs to be defined is open-textured, and how it should be applied varies with the context of each individual case. Another more general lesson learned is that the philosophical notions of definition and analogy that we started out with did not work very well as applied to the task of seeing how arguments from precedent are based on arguments from analogy. In particular, we see that the schemes for argument from analogy have to be reconfigured and fitted in to the schemes for argument from classification and argument from precedent to provide a basis for revealing how legal arguments from precedent work.

The most important conclusion of the chapter is that when similarity is defined in the way indicated using story schemes, we can reconfigure the argumentation schemes for argument from analogy in the way shown in Section 9, and thereby solve the problem of how they should fit together. The new version of the simple scheme functions as a device for identifying any given instance of an argument from analogy in a text of discourse. The identification requirements for the simple scheme can now be formulated as follows.

1. There is a database representing a set of source cases that have been decided.
2. There is common knowledge of one source case that is similar to the target case.

3. The source case has a particular (preferred) conclusion drawn from it.
4. The comparable outcome is argued for in the target case on the basis of the similarity.

This set of requirements helps the argument coder to identify an instance of argument from analogy found in a text.

The more complex scheme described by Guarini functions as a device for evaluating an argument from analogy as strong or weak. The idea is that the two schemes need to be employed in tandem, with the simple scheme being used first and the complex scheme being used to follow it up. This approach has better enabled us to bring out the significant relationships of argument from analogy with other closely related schemes such as argument from classification and argument from precedent. We are now in a position to see how the basic scheme ties in with the core scheme, and how the core scheme fits with tools such as use of factors and dimensions in case-based reasoning.

By carefully distinguishing these schemes from each other, and by contrasting them with related schemes, we were able to get a much more precise and useful theory that shows how argumentation from analogy works in case-based reasoning. In particular, we have seen how the notion of analogy needed to be reconfigured to provide a better basis for revealing how arguments from precedent work, based on a premise of similarity between two cases. In typical cases where an argument from precedent is used, as illustrated by the examples treated, it was seen to be based on an underlying argument from analogy. In dealing with arguments found in cases, there are two general sorts of tasks to be undertaken by argumentation methods. The first task is that we have to recognize arguments from analogy, and to do this we need to distinguish between it and other closely related arguments, such as argument from classification and argument from precedent. As the case of the drug-sniffing dog showed in Section 3, the first classification stage of reasoning by analogy leads to a later evaluation stage. The best device to be used for carrying out the first task is the argumentation scheme for argument from analogy, along with the matching critical questions and the other schemes studied in this chapter.

It was noted in Section 7 that the three-part argumentation procedure can be used to set up pre and post conditions for a dialogue model of use of argument from analogy, and the other related types of arguments we considered, using case-based reasoning. It was shown how Carneades can manage schemes and critical questions, and how the formal dialogue system of Bex and Prakken (2010) for investigating stories provides a framework for elucidating how the story-based model and the argument-based model of evidence can be combined in a unified formal framework. A future project is to show in more detail how the argumentation scheme for argument from analogy should be evaluated in these systems. Another project is to apply the argumentation schemes and the story scheme to further cases of argument from analogy. Such cases could be drawn both from law and from argumentation in everyday conversational discourse.

6

Teleological Argumentation to and from Motives

As shown in Chapter 1 Section 2, there are two apparently opposed models of rational thinking and acting in the literature on cognitive science. The belief-desire-intention (BDI) model is based on the concept of an agent that carries out practical reasoning premised on goals that represent the agent's intentions and incoming perceptions that update the agent's set of beliefs. The commitment model is based on agents interacting with each other in a dialogue in which each contributes speech acts. Commitment in dialogue is a public notion because evidence of commitment is available in the commitment set (database) of propositions that an agent has gone on record as accepting (Hamblin, 1970; 1971). Explicit commitments are statements externally accepted by an agent and recorded in an external memory that is transparent to all parties. Implicit commitments, of the kind that need to be postulated to reconstruct arguments as enthymemes, can be inferred from the explicit ones using argumentation schemes and dialogue rules, as indicated in Chapter 3. However, beliefs, desires and intentions are private psychological notions internal to an agent, and so there is a logical problem of how we are to infer what they are. One agent cannot directly inspect the contents of another agent's mind as a basis for making this judgment. This inability poses an evidential problem for the BDI model.

The problem of other minds has long been a central difficulty in philosophy, not only in philosophy of law, but also in ethics and philosophy of mind. We can observe a person's external actions, and thereby have empirical evidence to confirm or refute the claim that this person carried out a particular action. But since we cannot directly observe a person's motive, intention or desire, how can we confirm or refute any claim that a person acted on the basis of such an internal state of mind? The problem is particularly acute in legal reasoning about evidence in criminal law, because so much of it is built on assumptions about *mens rea*, the guilty mind. This chapter surveys recent developments in argumentation-based artificial intelligence and law to address the problem by studying the logical structure of reasoning about motives in law.

This chapter extends the theory of evidential reasoning about motives set forth by Walton and Schafer (2006), which provided a teleological framework for reasoning forward from motive to action, and reasoning backward from action to inferred motive. The extension of this earlier theory combines top-down and bottom-up models of teleological practical reasoning using argumentation schemes in a BDI model of practical reasoning. In this chapter, following the Walton and Schafer model, one intelligent agent reconstructs the motive of another agent by drawing an inference from facts and commitments of the other agent using abductive reasoning. Motives are defined as immediate internal desires to which an agent is strongly committed and has adopted as a mainspring of an action. However, another agent can reasonably infer that this first agent has a particular motive by using circumstantial evidence about the first agent's statements and actions.

It is shown in this chapter how argument visualization tools can be used to model such backward and forward reasoning by showing (1) how a case can be made for arguing that a motive led to an action and (2) how a motive can be attributed to an agent based on circumstantial evidence from his or her actions and speech. These findings are used to suggest that building an argumentation model of evidential reasoning from the facts of the case to a hypothesis about a motive in that case is an important first step to building a better BDI model that can help us reason about intentions and desires as components of practical reasoning. The findings of the chapter provide a bridge from the commitment model outlined in Chapter 1 to the BDI model as an evidence-based argumentation structure.

1. Some Short Examples and an Introductory Survey

Motive cannot be proved directly, because it is part of an agent's state of mind. It has to be proved indirectly by inference using circumstantial evidence (Wigmore, 1940, §§385, 327). Circumstantial evidence of an agent's motive comes from actions, either committed by the agent or by other agents, for example, from an injury that another party has done to the agent. Once the existence of the motive has been established, it can lead to a second inferential step (Leonard, 2001, 447) in which it is used to conclude that the agent committed a particular act, that the act in question occurred or that the agent had some state of mind (in criminal cases, a guilty mind).

In the following example (Leonard, 2001, 447), circumstantial evidence of the defendant's theft activity was taken as relevant evidence of his motive. The defendant was charged with the murder of the victim, but claimed not to have been involved. However, the prosecution had evidence that the defendant had been involved in a car theft prior to the killing, that the victim knew about the theft and that the victim had threatened to reveal the theft to the police. Leonard (448) structured the inference from evidence to motive in this case as follows.

Evidence: Defendant stole a car, victim was aware of the fact, and victim threatened to inform the police.

Inference: Defendant had a motive to prevent victim from revealing the theft to the police.

Conclusion: Defendant murdered victim to prevent victim from revealing the theft to the police.

Walton and Schafer (2006) showed how the reasoning used in this example is a combination of practical reasoning and abductive reasoning, or inference to the best explanation (IBE). IBE infers a conclusion from a set of observed or given facts or data by selecting the best one among several explanations that could account for the facts of a case. Typically, in such a case, the two sides have presented two opposed accounts, or stories, at trial, and IBE is used to point to the one as the better explanation (Bex, 2009a).

It has been shown by Pardo and Allen (2007) how the comparison of explanatory considerations can provide a better way of managing micro-level proof issues concerning the relevance and probative value of evidence in criminal cases. When the issue turns on two competing stories, it is necessary to go to a deeper level of analysis in which explanations are embedded within arguments. To analyze the deeper complexities of such evidential situations, it is shown in this chapter how attributing a motive to an agent needs a special type of teleological explanation based on what is called a story scheme (Bex, 2009b). Such an explanation is based on the factual evidence of the case, and therefore attributing a motive to an agent is also based on arguments. This structure is used to analyze inferences from a motive to an action, and from an action to an inferred motive.

Bex, Bench-Capon and Atkinson (2009) have also used the argumentation scheme for practical reasoning, along with a more extensive matching set of critical questions, and a technical apparatus called an action-based alternating transition system to model the evidential reasoning in a criminal case where one person was suspected of killing another person by pushing him off a bridge. Their analysis goes beyond the simpler one provided in this chapter by using a more fine-grained set of critical questions for choice of explanation. Their analysis is more technically powerful than the one presented here, and it shows several ways in which the simpler model presented here could be extended. However, it will be argued below that the simpler model also has some advantages as a representation of evidential reasoning about motives.

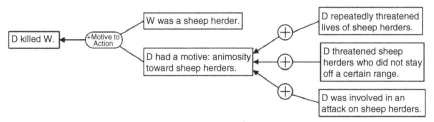

FIGURE 6.1 Argument Map of Argument from Motive to Action

Another example of argumentation from motive to action is the case of *Idaho v. Davis* (53 P. 678 Idaho 1898), which concerned the struggle between sheep herders and cattlemen to control land. The prosecution offered three pieces of evidence to prove that D, a cattleman, killed W, a sheep herder, shown in the three text boxes on the right in Figure 6.1. The plus in the node denotes a pro argument, an argument that provides positive support for its conclusion. The argument shown in Figure 6.1 presents the standard method of argument diagramming where the text boxes and containing propositions are used as premises or conclusions in an argument. This particular argument map was drawn using the Carneades Argumentation System. In this system, the statements in the text boxes can be designated as questioned, stated, accepted or rejected. Once the premises of the argument are evaluated in this four-valued system, and the arguments are configured into structures using argumentation schemes so that each argument is applicable, Carneades automatically designates one of the four values to the conclusion.

The argument shown in the map in Figure 6.1, drawn using the graphical user interface for the Carneades Argumentation System, concludes in an action drawn from the postulation of a motive, which is in turn supported by an argument from factual evidence. As shown in Chapter 2, the Carneades Argumentation System uses argumentation schemes and models the critical questions matching a scheme as premises of the scheme (Gordon, Prakken and Walton, 2007).

The part of the argument comprised of the two premises and the conclusion shown at the left of Figure 6.1 is an instance of argument from motive to action. The argumentation scheme for argument from motive to action is formulated below, based on the comparable form of inference described by Leonard (2001, 59).

Conditional Premise: If agent *a* had a motive to bring about *A*, then *a* is somewhat more likely to have brought about *A* than another agent who lacked a motive.

Motive Premise: *a* had a motive to bring about *A*.

Conclusion: *a* brought about *A*.

This form of inference, as structured by Leonard, with two premises and a conclusion, can be modeled as an argumentation scheme for argument from motive to action. So far, then, we have seen how an argument that goes from a motive to an action can be configured with this argumentation scheme. But we are still left with the problem of how to argue the other way around, from facts about actions and circumstances to a motive. Teleological reasoning can be used to establish the existence of a motive by drawing an inference from premises concerning facts of a case to a conclusion that a motive exists (Walton and Schafer, 2006). A sequence of teleological reasoning leads from a set of circumstances in a case to a hypothesis that postulates the existence of a motive. To see how this was done in Walton and Schafer (2006), we have to put this scheme into a broader argumentation framework by recalling the argumentation scheme for practical reasoning in Chapter 1, Section 6.

In this scheme, the first-person pronoun 'I' represents a rational agent that has goals, some (though possibly incomplete) knowledge of its circumstances, the capability of acting to alter those circumstances and the capability of perceiving the consequences of acting.

Major Premise: I have a goal *G*.

Minor Premise: Carrying out this action *A* is a means to realize *G*.

Conclusion: Therefore, I ought (practically speaking) to carry out this action *A*.

Here once again from Chapter 1, Section 6, are the five critical questions matching this scheme

CQ$_1$: What other goals do I have that should be considered that might conflict with *G*?

CQ$_2$: What alternative actions to my bringing about *A* should also be considered?

CQ$_3$: Among bringing about *A* and these alternative actions, which is the most efficient?

CQ$_4$: What grounds are there for arguing that it is practically possible for me to do *A*?

CQ$_5$: What consequences of my bringing about *A* should also be taken into account?

The last critical question, CQ$_5$, is very often called the side effects question. It often concerns potential negative consequences of a proposed course of actions. For a more complex and powerful system of value-based practical reasoning, see Atkinson, Bench-Capon and McBurney (2005). Practical reasoning can move forward, from a goal to an action, as part of agent-based deliberation, but it can also be used backward, to reconstruct a plausible motive based on an agent's actions and words. When used in this backward fashion, we begin with the conclusion and the means premise and reason to the hypothesis that the agent acted on a goal that would fit the actions into a practical reasoning structure that would provide a rationale for acting on a goal.

2. Relevance of Motive Evidence and Character Evidence

The Federal Rules of Evidence allow character evidence to be admissible only in certain instances.[1] Rule 401 defines relevant evidence as evidence having any tendency to make the existence of any fact "that is of consequence to the determination of the action more probable or less probable than it would be without the evidence". Relevance according to the account given in the Federal Rules of Evidence is defined in terms of what is called probative weight or probative value. An argument is admissible as relevant in a trial only if it makes the ultimate proposition to be proved more probable or less probable. However, 'probability' is to be understood here not in the narrower sense of statistical probability but in a broader sense meaning that factual evidence can combine with logical reasoning to make a conclusion carry more or less probative weight than it did without the evidence.

According to Rule 403, relevant evidence (according to the requirements of Rule 401), may be excluded if its probative value "is substantially outweighed by the danger of unfair prejudice, confusion of the issues, or misleading the jury, or by considerations of undue delay, waste of time, or needless presentation of cumulative evidence". Even if evidence is relevant, it may be inadmissible if it might tend to prejudice a jury. The worry about character evidence expressed by Rule 403 is that it may be too persuasive in its leading a jury to give it too much weight. Character attack, in the form "He is a bad person, therefore he must be guilty", is a powerful form of argument, because we do not trust people who are thought to have committed crimes. The character attack form of argument, called the *ad hominem* argument in logic, which is basically refuting someone's argument by attacking their character, is sometimes reasonable but can often have so much undue impact on an audience that it has traditionally been considered to be fallacious (Walton, 2006b).

Rule 404 states that character evidence is not admissible for the purpose of proving conduct. Rule 404(b) says that evidence of other crimes, wrongs or acts is not admissible to prove the character of a person in order to show action. But there are exceptions. One is that if character evidence is introduced by the defense, the prosecution can then use character evidence in rebuttal. Another is that character evidence can be used in examining witness testimony. It can also be used if character is an essential element of a charge or defense. For example, character of a person would be relevant to the issue of whether the defendant was negligent in hiring or entrusting property to an unfit person. Evidence of crimes or bad acts may also be admitted if the evidence is offered, not to show character, but for some nar-

---

[1] The latest version of these rules can be found on the Web at www.uscourts.gov/rules/newrules4.html.

rower purpose such as showing motive, opportunity, intention, preparation, plan, knowledge, identity or absence of mistake or accident.

The problem introduced now is how to distinguish between motive and character. However, since this task has already been commented on and dealt with in Golden (1994), Leonard (2001), Sartor (2005) and Walton (2006b), it will be set aside here. The other large problem set aside for further work is the difference between motive and intent. Leonard (2001) is very helpful here, even providing an argument scheme for argument from motive to intent.

3. Inference to the Best Explanation

Abductive reasoning, very important for scientific discovery, is here equated with IBE. Abductive inference fits the following format (Josephson and Josephson, 1994), showing its structure as inference to the best explanation.

- H is a hypothesis.
- D is a collection of data.
- H explains D.
- No other hypothesis can explain D as well as H does.
- Therefore H is probably true.

In inference to the best explanation, multiple explanations are generated and comparatively evaluated according to criteria that express the degree to which they conform to the evidence and their plausibility. Explanations are evaluated by means of arguments, and so it is clear that argument and explanation are closely interwoven.

1. Arguments based on evidence can be used to show that an explanation is consistent or inconsistent with the evidence.
2. Arguments may also be used to reason about the plausibility of an explanation, as the validity and applicability of causal rules can become the subject of an argumentation process.
3. Arguments about the plausibility of explanations are based on plausible reasoning, carried out by using commonsense knowledge about how the world generally works in familiar situations.

One of the key questions in analyzing abductive reasoning as IBE is to analyze the notion of explanation on which it depends.

The main problem with modeling abductive reasoning (IBE) is to furnish an analysis of the concept of explanation that is better than the traditional deductive nomological model that held sway for so long in analytical philosophy. This project has been carried forward in Walton (2011a) by building a dialogue system of explanation with rules that define kinds of speech acts appropriate for asking for and offering an explanation. In this dialogue model, a successful explanation has been achieved when there has been a transfer of understanding from the party offering the explanation to the party requesting it. As far as inference to the best explanation is concerned, the problem is to have criteria that enable us to determine when one explanation is better than another, and, given a set of competing explanations, to determine which is the best. In this dialogue-based theory, a crucial role is played by the notion of a script or story, because an explanation is determined by how a story fits together and how well it stands up to critical questioning. The story is said to be plausible if it not only fits together but is supported by relevant evidence and can also survive the process of testing through critical examination of how it fits together.

There are several steps to the dialogue procedure in this model of explanation. First, both parties have to begin with a coherent story that represents an account of an event they can understand, even though there will be parts of it that one party understands better than the other. When one of the parties finds an anomaly, something he or she does not understand, he or she asks a question requesting an explanation by the other party. Either the explanation is successful in transferring understanding to the party asking the question or it is not. If it is successful, the dialogue stops there. If it is not successful, the dialogue continues with further questions being asked and answered. In some cases there will be a shift to an examination dialogue where the explanation is tested by the questioner's critically probing into the weaknesses in it. Such a probing process may turn up inconsistencies, statements that are implausible or other kinds of anomalies, for example, gaps in the story. How good an explanation is judged to be in a dialogue is partly determined by how it can stand up to such questioning.

Practical reasoning is based on common knowledge; it is defeasible; it is based on the way things generally go in familiar situations; it can be used to fill in implicit premises in incomplete arguments; it is commonly based on appearances (perception); it can be tested against facts and is by this means confirmed or refuted. Probing into practical reasoning in a critical examination is a way of testing it.

An example from Wigmore (1940, 420) shows how he analyzed cases of legal evidence as instances of inference to the best explanation.

The fact that *a* before a robbery had no money, but after had a large sum, is offered to indicate that he by robbery became possessed of the large sum of money. There are several other possible explanations – the receipt of a legacy, the payment of a debt, the winning of a gambling game, and the like. Nevertheless, the desired explanation rises, among other explanations, to a fair degree of plausibility, and the evidence is received.

The evidence put forward in this example has the form of inference to the best explanation. It shows the conclusion as arrived at by means of a choice among several competing explanations of the given facts.

The argumentation scheme for an abductive argument is based on two variables: the variable F stands for a set of facts, and the variable E stands for an explanation. The concept of explanation is dialectical. An explanation is a response to a question in a sequence of dialogue. Below is the argumentation scheme for abductive argument (Walton, 2006b, 167), comparable in structure to the Josephsons' model.

Facts Premise: F is a finding or given set of facts.

Explanation Premise: E is a satisfactory explanation of F.

Alternative Premise: No alternative explanation E' given so far is as satisfactory as E.

Conclusion: E is plausible as a hypothesis.

This form of argument is defeasible. It can be defeated by asking appropriate critical questions.

CQ_1: How satisfactory is E itself as an explanation of F?

CQ_2: How much better an explanation is E than the alternative explanations available?

CQ_3: How far has the dialogue progressed?

CQ_4: Would it be better to continue the dialogue, instead of drawing a conclusion now?

This scheme is dialectical, meaning that it is evaluated in a dialogue in which one puts forward a conclusion based on an argument, and the other party asks critical questions or puts forward counterarguments that may defeat the argument (Prakken and Sartor, 2006a).

4. Stories and Explanations

In research on reasoning with criminal evidence, two main trends are the argumentation approach and the narrative approach. Arguments are constructed by taking items of evidence and reasoning toward a conclusion respecting facts at issue in the case. It has been characterized as evidential reasoning because of the relations underlying each reasoning step: 'a witness testifying to some event is evidence for the occurrence of the event'. Hypothetical stories based on the evidence can be constructed, telling us what might have happened in a case. Alternative stories about what happened before, during and after the crime can then be compared according to their plausibility and the amount of evidence they explain.

The notion of a story as an account of some event based on a so-called script was explained in Chapter 5, Section 5. On the logical argumentation model outlined in Chapter 1, a story is a set of statements offered by one party in a dialogue in answer to questions put by the other. A story is set of statements linked to each other by a series of relations connecting an agent's goals to his or her actions. A story does not have to be internally consistent, but if an inconsistency is found, questions can be asked, and the story might have to be repaired or given up. If one of a pair of competing stories is more plausible, all else being equal, the more plausible one should be accepted as the better explanation. The dialogue process of examining a story starts with a database representing the facts so far collected in an account. Examination is a complex process that typically begins with an explanation but can shift to a critiquing phase in which the story in it is probed for questionable gaps and apparent inconsistencies. The questioner asks a question to achieve a better understanding of some or all of these facts. The respondent replies by putting forward a story offered to explain the facts that were asked about. Alternative stories that serve to explain the same facts may also be given. The comparative plausibility of each story is judged by how well each stands up to critical questioning.

There are seven factors that can be used to judge how good a given story is as an explanation compared with another story: (1) how well it performs its function of helping a questioner to make sense of something, (2) whether it is internally consistent or not, (3) whether an alleged inconsistency can be dealt with, (4) how well it is supported by the factual evidence, (5) how plausible the account is generally, (6) how comprehensive and detailed it is in covering relevant events and actions and (7) how well it stands up to critical questioning and examination.

Pennington and Hastie (1993) showed how actions carried out in criminal cases can be explained by competing explanations in which each provides a story connecting the facts into a sequence that seems plausible based on the evidence. They argued that understanding actions carried out in criminal cases is done by constructing competing stories about what supposedly happened using the evidence in the case. Such a plausible story describes a general pattern of actions and events of kinds with which we are all familiar. One story can be more plausible than another. However, a plausible story may not be very well supported by the evidence, whereas a less plausible story may be supported by more evidence. To solve this problem, Wagenaar, van Koppen and Crombag (1993) devised a special type of story used to represent legal reasoning called an anchored narrative. As shown in Chapter 5, Bex, (2009a; 2009b) proposed a hybrid framework for reasoning with arguments, stories and criminal evidence, a formal framework that shows how the plausibility of the story can be evaluated by giving arguments that ground the story on evidence that supports or attacks it.

Bex (2011) modeled a story as an ordered list of events or types of events that can be more abstract or more specific. In the example outlined in Chapter 5, Section 5 (Bex, 2009b, 59), John Haaknat was a drug addict who needed money and decided to rob a supermarket. He got the money, jumped into his car and sped away, but then he saw the police, and parked his car in a park. He then abandoned the car and jumped into a moat to hide. When the police searched the park they found him soaking wet from water in the moat. Bex (2011, 59) showed how a visual representation of the story can display the ordered structure of the events in it, using an explanation diagram like the one shown in Figure 5.2.

The story visually represented in Figure 5.2 shows how the actions of Haaknat are combined with what are taken to be his mental states. Because he is a drug addict who needs money, we can infer that he decided to rob the market to get the money. After he robbed the market and took off in his car, we can infer that he went to the park to hide because he thought the police were after him. From his actions and from evidence of what he has said, we can draw plausible conclusions about his mental states and how they fitted in with the sequence of actions that he carried out.

Haaknat was found hiding in a moat in the park after the robbery, and the prosecution's explanation was that he had fled there after the robbery to avoid arrest. Haaknat's explanation was that he was hiding in the moat because he had an argument with a man over some money, this man had drawn a knife, and he had fled to escape this man. Here we have two competing stories; the problem is to try to judge which is the more plausible, based on the facts.

Bex and Prakken (2010, 5) showed how two competing explanations in a criminal case can be evaluated by criteria to judge which is the better. One criterion they use is evidential coverage, meaning how many arguments can be used to support claims that are parts of the explanation. Another is the internal consistency of each story. Bex and Prakken provide a formal dialogue model that can be used to evaluate the arguments on each side, and pose critical questions to test the plausibility of a story. Bex, Prakken, Reed and Walton (2003) showed how evidential reasoning in law is typically based on general knowledge accepted in a community codified in defeasible generalizations.

As outlined in Chapter 5, Section 6 (Bex, 2009b, 126), a story scheme is defined as a collection of propositions and generalizations with a set of inference rules for classical logic with a defeasible *modus ponens* rule for a conditional operator ⇒ that represents defeasible generalizations (Bex and Prakken, 2010). Story schemes divide the states and events in a story into different categories, such as actions and relations between actions. Story schemes are abstract representations of stories of the kind represented by scripts. For example, in the bank robbery story scheme, one party is taking some money or valuable goods from another party, the person being robbed (Bex 2009b, 64). There is a motive for the robbery; the robber acts from a motive; the motive is to acquire some goods possessed by the party being robbed; there is force employed by the robber; the robber has an opportunity to take these goods from the party being robbed; and the party who is robbed loses the goods. Story schemes can be compared to argumentation schemes. An argumentation scheme is a general scheme for arguments of a particular type, and in a comparable way the story scheme is a general scheme for stories of a particular type (Bex 2009b, 65). Story schemes divide the actions and events in the story into different categories, such as the category of the action of robbery or the category of the participant of robber. They also define relationships between these categories, showing, for example, how one action or event is related sequentially to other actions or events.

5. Who Shot the Sheriff?

In the case of *State v. Brown* (398 So. 2d 1381 (La. 1981)), the defendant B was charged with attempted first-degree murder of a deputy sheriff who had stopped the defendant's car for speeding. A car driven by B, with W in the front passenger's seat, was stopped for speeding by Deputy Sheriff G. G got out of his car and walked toward the stopped car, but as he reached a point close to the rear of the car, he saw the defendant pointing a shotgun through the car window. It appeared to him that B was trying to fire the gun at him, but it had misfired, so he turned and ran away, and was shot in the shoulder from behind (later, G died). He then jumped into a ditch and fired six times with his revolver as the car sped away. B and W abandoned the car after a short distance and tried to escape by running across a levee, where they were apprehended by police.

B later testified that W, who owned the gun, handed it to B and told him to shoot G. He also testified that W later took the gun and shot G and that W told him to leave the scene after the shooting. Later testimony of a used car dealer presented clear and convincing evidence that B had stolen the car. At issue was whether this evidence was admissible. As character evidence it would not be admissible; however, as motive evidence, it could be used as part of the evidence to prove that B had committed the crime of second-degree murder.

The problem was whether evidence that the defendant had stolen the car was admissible as an exception to the general rule barring admissibility of evidence of previous crimes. If this evidence was being used to show that the defendant had a motive for shooting the sheriff, it could be considered relevant. The link between such a motive and the shooting was drawn by the court in the following words: "If defendant had stolen the automobile, a crime for which he could be sent to prison for many years, it was most important for him to avoid having the crime discovered, a very likely probability in the event he was arrested on the speeding charge". The court admitted the evidence that the defendant had stolen the vehicle on the grounds that it established a motive for him to fire on a deputy sheriff in order to avoid being arrested on the speeding charge.

To analyze the evidential reasoning in this case, we have to go back to try to reconstruct the state of mind of the defendant when his car was stopped for speeding. The consequences of being given a speeding ticket are not too serious, probably paying a fine. However, the consequences of stealing

FIGURE 6.2 Structure of Practical Reasoning in the Sheriff Example

a car are likely to be much more serious in comparison. As the court stated, it is a crime for which the defendant could be sent to prison for many years. The defendant, we may presume, knew about these probable consequences and their comparative seriousness. When making a decision on what to do, taking these negative consequences into account, it is plausible that he acted impulsively to avoid the more serious negative consequences, but there was a kind of practical reasoning involved (Figure 6.2).

It almost seems inappropriate to call avoiding arrest for car theft a goal, because that way of classifying it seems to suggest a kind of foresight, planning or rational calculation that perhaps should not be attributed to an action that appeared to be impulsive. But if we can classify a motive as a goal, something that provides a mainspring of action and therefore leads to action by prompting action immediately, we could replace the goal text box in Figure 6.1 with a box containing a motive. We don't have a name for this motive, but it could be called 'discovery avoidance'. Discovery avoidance could be a lively mainspring for action, and hence could act as a motive. Shooting a deputy sheriff is an action that is also very likely to have serious negative consequences, even going to prison for a longer time, so if we saw the motive as acting as a kind of goal, we could see it as fitting the practical reasoning scheme along with its critical questions.

So far we are just at the start point of the analysis of this case by trying to structure it as motivated action in the mind of the agent at the time he carried it out. Next, we need to look at it from the point of view of the court trying to connect motives to an action at issue in the trial. To do this, we represent it as an instance of argument from motive to action using Carneades. We can say that the example in this case is a straightforward application of the argumentation system. Suppose, for example, that all three of the premises on the right are designated as accepted. In the system this would be shown by placing a check mark in front of the proposition in each of the three boxes. Then the system automatically puts a check mark in the middle box, and a check mark also in the text box on the extreme left. As shown in Figure 6.3, all the text boxes contain check marks, and each text box is darkened as well, showing redundantly that the statement in the text box has been accepted.

The argument shown in Figure 6.3 has three premises at the right that are linked together to support the proposition that B had a motive to shoot

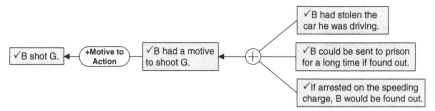

FIGURE 6.3 Argument from Motive to Action in the Sheriff Case

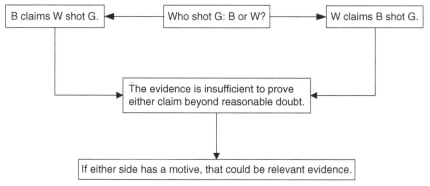

FIGURE 6.4 Two Competing Claims on a Balance

G. This proposition in turn leads to the conclusion that B shot G. This figure shows how the circumstantial evidence supports the conclusion that there was a motive as the basis for an action.

A special aspect that is important is relevance. What made the argument from motive to action relevant is that there was a balance in the case and insufficient evidence to resolve the issue (Figure 6.4). The argumentation is on a balance, because each party claims the other shot G and there was no circumstantial evidence or witness testimony to show which party was the shooter.

There are two claims or stories. There is no circumstantial evidence that supports the one story more strongly than the other that could be used to tilt the balance or burden of proof to one side or the other. Since G's back was turned when he was shot, he could not see which man in the car fired the shot. Moreover, the conclusion at issue of who shot the sheriff has to be proved to be beyond reasonable doubt. The problem is to see how the motive evidence can be factored in to tilt the balance to one side or the other.

The balance situation can be represented better by comparing two argument maps representing the evidence on each side. This is shown in Figure 6.5. At the top of Figure 6.5 the evidence supporting the conclusion that B shot G is displayed. At the bottom of the figure, the lack of evidence supporting the opposed claim that W shot G is shown.

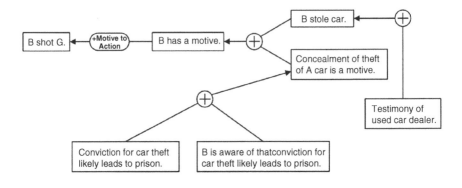

FIGURE 6.5 Argument Maps for the Evidence on Each Side

When trying to reconstruct a motive, it is not a matter of balancing values and probabilities to determine what is or was a rational course of action. Rather, it is a matter of reconstructing the line of plausible reasoning of the agent so that we can understand what gave him a reason to act in a certain way, even though we think that the way he actually acted was irrational, unethical, illegal or otherwise subject to condemnation or criticism.

So far, our analysis of the example seems superficial, and perhaps even not very convincing, because, as we will now show, it looks only at the surface of the argumentation. To get further, we have to bring in explanation as well as argumentation and see how the two can be combined to produce an IBE model of the evidential reasoning.

6. Going from Argument to Explanation to Motive

We can get a deeper appreciation of evidential reasoning about motives if we look beneath the layer of argumentation to the layer of explanation that lies underneath it. From an explanation point of view, you need to look at how the example begins with an anomaly, an unusual or puzzling set of circumstances that calls for an explanation. This first stage of the analysis is shown above the line labeled 'explanation' in Figure 6.6. Once the anomaly is posed, the explanation below the line in Figure 6.6 purports to resolve it. The structure represented in Figure 6.6 shows a sequence of events represented as propositions in the nodes joined by arrows that represent transitions from one node to the next.

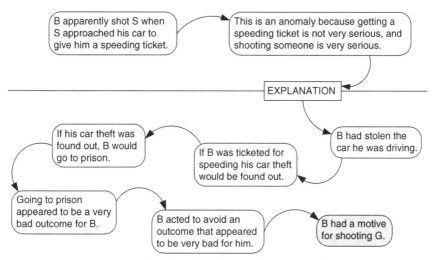

Figure 6.6 An Explanation Leading to a Motive

The part of Figure 6.6 below the line labeled 'explanation' is a story that links several events together into a sequence so that we can understand it as a coherent whole because of our common knowledge of how actions are connected in a script. Pennington and Hastie (1993) devised the notion of an episode scheme that can be seen as a script that describes the general pattern of events like those in the famous restaurant script. Bex (2009b) combined the episode schemes of Pennington and Hastie with the scripts of Schank (1986) and his colleagues (Schank and Abelson, 1977) to form what are called story schemes.

These are modeled as an ordered list of events or types of events that can be more abstract or more specific. Bex (2009b, 59) used the example shown in Figure 5.2, where Haaknat is a drug addict who needs money and decides to rob a supermarket. Haaknat carries out other actions that fit into an episode sequence that includes his jumping into a moat. Later the police search the park and find him soaking wet from the water in the moat. The sequence shown in Figure 6.6 is a story scheme of the kind shown earlier in Figure 5.2. However, it is a special kind of story scheme that represents the structure of an explanation starting from an anomaly and ending in a motive. Let's call it an IBE motive story scheme. In Figure 6.6, the anomaly to be explained is shown above the dividing line, while the episode sequence below the line shows the explanation of the anomaly.

To complete the analysis of the role of motives in the evidential reasoning in the case of who shot the sheriff, we also have to look at the story scheme on the other side. The evaluation of the evidence in a typical case, according to the theory presented here, proceeds by examining the

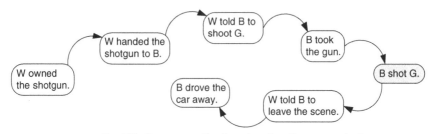

FIGURE 6.7 W's Story as an Explanation Leading to an Action

stories on both sides and deciding which is the more plausible or best explanation of the facts in the case. B's story was that W shot G, but W's story was that B shot G, visually represented in the episode scheme shown in Figure 6.7.

By using story schemes to represent the competing explanations on both sides, we have been able to give a deeper analysis of the evidential reasoning in the case of who shot the sheriff.

Bex, Bench-Capon and Atkinson (2009, 96) provided an even deeper analysis of Leonard's car theft case (Section 1) that not only takes into account reasoning backward from the circumstances in this case to the motive, but also considers motives for several alternatives. Their analysis takes into account better ways for the defendant of achieving his goal of not being punished. The alternatives they consider are paying the victim to keep silent or leaving the country. They use a technical system called a value-based argumentation framework to look for evidence that tells a story to explain why the defendant is the sort of person who could not afford to pay the victim, who values life cheaply, who accepted the risk of getting caught for murder and who preferred killing the victim to leaving the country. This technical system is so powerful that it could be used as a method of argument invention to tell a lawyer searching for evidence where to look next. However, it also has implications for admissibility of evidence because it can show us where evidence of character and past actions become relevant.

7. Matters of Argument Evaluation

The focus of this chapter is been on admissibility of the motive evidence in the case of *State v. Brown*. However, there is an issue concerning the evaluation of the evidence in the case that raises a problem. As shown in Figure 6.4, the evidence in the case is insufficient to prove either of the two competing claims to the standard of beyond a reasonable doubt. The charge in this case was that of attempted first-degree murder, and because it is a criminal case, to convict the defendant the trier must prove the charge beyond reasonable doubt. Examining the motive evidence in the case using the analysis above, it looks very much like this evidence is insufficient to meet the beyond reasonable doubt standard of proof. However, when the evidence is sufficient or does not meet that standard may depend on how the beyond reasonable doubt standard is defined. Standards of proof are built in burdens of proof.

Prakken and Sartor (2009, 228) have built a logical model of burden of proof in law. The burden of persuasion specifies which party has to prove some proposition that represents the ultimate *probandum* in the case, and also specifies what proof standard has to be met. The burden of persuasion is set at the opening stage of a trial and is a global burden that applies through the whole sequence of argumentation to the closing stage, where it is used to determine the winner and loser of the case. The burden of production specifies which party has to offer evidence on some specific issue that arises during a particular point during the argumentation stage of the trial. Both the burden of persuasion and the burden of production are assigned by law. The tactical burden of proof is a determination made by the advocate in building a sequence of argumentation and is an estimate of whether his or her present argument will fail if he or she fails to support it further. This is not set by law but is only an estimate made by the arguer. Only the tactical burden of proof can shift back and forth from one party to the other (Prakken and Sartor, 2009). The beyond reasonable doubt standard represents burden of persuasion in a criminal case.

It is generally regarded as very dangerous for a judge to try to define 'beyond reasonable doubt' in a criminal trial, as there is judicial hostility to attempting any precise definition, and there is a very real danger of appeal, because such a definition is not established in precedent (Tillers and Gottfried, 2006).There is a vocal acceptance in law of the view that the beyond reasonable doubt standard cannot be quantified by using numbers. That does not mean it cannot be modeled using computational tools. Pardo and Allen (2007, 238) argue that it can modeled using inference to the best explanation. On their view, in criminal cases, rather than inferring the best explanation from the potential ones, fact-finders ought to convict when there is no plausible explanation consistent with innocence, assuming there is a plausible explanation consistent with guilt. On this model, a plausible explanation for innocence creates reasonable doubt, while a plausible explanation can be a basis for conviction on the beyond reasonable doubt standard if every competing explanation is so weak that it fails to raise a reasonable doubt. These circumstances include the case where there is only one plausible explanation, and there are no competing explanations that have been offered.

Standards of proof are formalized in the hybrid theory. Definitions of standards of proof formulated in the Bex and Walton (2010) model are set on a basis of how much better one explanation is than another, and how good an explanation is in itself. Following Gordon and Walton (2009), standards of proof are not given fixed values but left open to be set by an argument evaluator when the model is applied to a case. An explanation *EX* is said to meet the scintilla of evidence (SE) standard if there is a supporting argument based on evidence ($es(EX) \geq 1$). An explanation *EX* meets the preponderance of evidence (PE) standard if it meets the SE standard and it is better than each alternative explanation *EX'*. In other words, all else being equal *EX* is either supported by more evidence ($es(EX) > es(EX')$) or contradicted by less evidence ($ec(EX) < ec(EX')$). For clear and convincing evidence (CCE), an explanation *EX* should be good in itself as well as much better than each competing explanation *EX'*, meaning it should have a high evidential support and low evidential contradiction. Finally, an explanation can meet the beyond-reasonable-doubt standard if it is plausible and each of its competing explanations is implausible. How plausible it needs to be, or how implausible the competitors need to be, is left open, but it is assumed that it needs to be highly plausible, and if there are competitors, they need to be not plausible at all. But precise evaluations of how plausible an explanation needs to be, and how much evidence there needs to be supporting it, is not specified numerically in the model. This approach provides a way of modeling standards of proof based on inference to the best explanation in such a way that burdens of proof can be determined by evaluating how plausible an explanation is in relation to competing explanations that can be given in a particular case by using the factual evidence in that case.

This approach is different from the usual one, in that the usual or standard approach to the evaluation of evidence is to consider a chain of argumentation that supports or attacks an ultimate *probandum* that is at issue in case and to see how strongly the premises and conclusions in that chain are supported by the evidence. A burden of persuasion has to be set for the ultimate *probandum* and burdens of production and tactical burdens of proof have to be set for the individual arguments in the chain. The hybrid approach is different. It looks at the stories given by both sides to explain the facts of a given case and evaluates these stories as more plausible or less plausible than competing stories. But in addition to the plausibility and coherence of explanations, it also takes into account how well the factual evidence supports or detracts from the story on each side.

Using this approach, the evidential reasoning about motives in the case of *Brown v. Alabama* can be evaluated in relation to the comparative explanations of what supposedly happened, as shown in Figures 6.4–6.7. Given all the circumstantial evidence in the case, and the testimony of the deputy sheriff, it is known beyond a reasonable doubt that one or the other of the two suspects shot the sheriff. However, as shown in the balance represented in Figure 6.4, this circumstantial evidence does not tilt decisively to one side or the other. On the one side, we have a story that is more plausible because it is supported by motive evidence. On the other side, we have a story that is less plausible because there is no comparable motive evidence supporting it. The problem then is whether the motive evidence should be strong enough as a basis for convicting the defendant using the beyond reasonable doubt standard. The analysis of motive evidence presented in this chapter requires that there be a sufficiently plausible explanation on one side, and a sufficiently implausible explanation on the other side, in order to prove the proposition that B shot the sheriff on the beyond reasonable doubt standard of proof. Whether the case was decided on this or some other basis is a matter for speculation.

In any event, enough has been shown here to see how the beyond reasonable doubt standard is modeled by the hybrid theory in a way different from the usual approach. The usual approach, as indicated above, takes into account only the evidence supporting the arguments and the proof standard the chain of argumentation has to meet in order to prove the conclusion at the end of it. The hybrid approach works differently, by evaluating the plausibility of competing explanations but taking into account how each competing explanation is more strongly or weakly supported by the evidential facts in a case. In the case of *State v. Brown*, there was a good deal of evidence on both sides, and since the motive evidence was admissible, it acted as a tiebreaker.

Finally, we return to the question of how the two explanations in the case of *State v. Brown* can be visualized on the hybrid model. We recall that in Figure 6.6 the story that explains B's presumed motive for shooting the sheriff was represented as an episode scheme. The problem now is how to factor in evidence that was available in the case that could be used to support the plausibility of this explanation. There is some explicit evidence given in the court summary of the case, but if we were to look at an expanded account of the transcript of the court proceeding, there might be other evidence as well that could be taken into account. Also, there is other implicit evidence in the form of implicit premises that are always significant in order to piece together the sequence of events using scripts to build stories that make sense. For example, one of the assumptions in the explanation is that B going to prison would be a very bad thing for him. How do we know this? There is no explicit evidence given, but it seems a reasonable assumption based on common knowledge that once somebody has been to prison, they realize that it is an experience they would care to avoid in the future if possible. As noted in the discussion of Figure 6.7, an episode scheme describes the general pattern, like those in the restaurant script, based on common knowledge about the way we expect things normally to go in familiar situations.

In Figure 6.8, none of these implicit premises has been represented. Only the explicit items of knowledge extracted from the details of the case

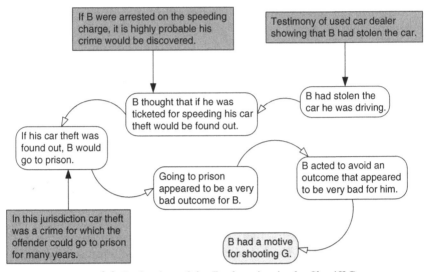

FIGURE 6.8 Evaluation of the Explanation in the Sheriff Case

known from the description of it have been inserted as items of evidence that support parts of the explanation. In Figure 6.7, the story is represented as a sequence of rounded boxes with no shading. The propositions that appear in the rectangular boxes that have shading in Figure 6.8 represent items of evidence. The lines with closed arrowheads, leading from a rectangular box, represent arguments from evidence. The lines with open arrowheads connecting the rounded boxes represent explanatory connections.

As the hybrid diagram in Figure 6.8 shows, there is evidence supporting the story that explains B's motives for shooting G. Evaluating how good the story is as an explanation of the facts of the case depends on the seven factors cited in Section 4. First, we have to ask how well the story performs its function of helping the questioner makes sense of the facts that were known to have taken place. The story appears to perform well in this regard. It makes sense as a coherent whole that we can understand. Second, it is internally consistent. Third, there appear to be no inconsistencies among the components that make it up, including both the evidential boxes and the story boxes and their relationships. Fourth, since there are no inconsistencies, the criterion that any alleged inconsistency can be dealt with is satisfied. Fifth, the story is generally a plausible account. Sixth, it appears to be comprehensive and detailed enough to account for the events and actions known in the case. Seventh, we don't know how well it stood up during the trial to critical questioning on examination. However, as far as we know from the court summary, no problems were found in the story during cross-examination at trial.

8. Wigmore's Theory

Wigmore (1931, 146) analyzed motives as emotions that lead to action. Examples he gave are desire for money that led to the action of robbery, and angry hostility that led to a violent criminal act. On his view, evidence of motive is usually circumstantial, consisting of the conduct of the person and the events in the particular situation tending to excite the emotion. He added (147) that there is an unfortunate ambiguity in the word 'motive', illustrating it with an example in which a defendant is accused of burning down a plaintiff's house. The plaintiff contends that the defendant's motive for burning down his house is his prior prosecution of a lawsuit against the defendant. Writers would sometimes cite the external fact of the plaintiff's lawsuit as the motive. But there is another description of that is very common as well. They might also, more properly (Wigmore, 1931, 147) refer to the defendant's hostile and vindictive emotion arising from the lawsuit as the motive. Wigmore allows that the term 'motive' may have either meaning. It can be a particular event or fact (set of circumstances) known to an agent, or it can be the lively emotion of that agent arising from that event.

In some cases, we don't have any description of the particular circumstances that constitute motive, even though we have evidence of an agent's past actions that support the existence of the motive. For example, in the sheep herders and cattlemen case, we know some past actions of the defendant, such as threatening the lives of sheep herders, but the only way we can describe the motive is an animosity toward sheep herders. If we have a known set of circumstances that led to the agent's action, we can use the model of practical reasoning described in Walton and Schafer (2006).

In cases where we have only the motive described as an emotion, a different model of practical reasoning model works better. In this model practical reasoning takes the following form: I (an agent) desire to achieve end E; I believe that the best way to achieve E is to do M; therefore I do M (Searle, 2001, 244–246). The major premise in the commitment-based argumentation scheme is a statement that the agent has a goal, whereas the major premise in the BDI scheme is the statement that the agent has a desire or want. Desires (wants) are deeply psychological entities, more so than goals. Goals are meant in the standard argumentation scheme for practical reasoning to be statements designated as ultimate points in an agent's plan toward which the actions contemplated in the plan are directed to bringing about. Desires represent an agent's emotional attachment to an object or state of affairs that the agent wants to bring about and may even be strongly impelled to bring about, although he may resist that impulse.

In the literature on practical reasoning, the BDI model tends to be dominant, but there is also a minority who advocate the commitment-based model. It has been perceived as a problem with the BDI model that because it expresses the premises of the practical inference in terms of beliefs and desires, it is too psychological in nature to make the practical inference rationally binding (Walton, 1990, 26–31). Also, the commitment-based model fits very well into the logical structure of the technology of planning, a well-developed branch of artificial intelligence.

A nice feature of the BDI model, however, is that precisely because of its deeply psychological nature, it fits extremely well into the study of how emotions can act as motivating factors in deliberations on how to carry out actions. Wigmore (1931, 147) shows that despite the ambiguity in the word 'motive', for use in legal reasoning he prefers the meaning that motive should be a state of mind as opposed to an external fact. He wrote (147): "that which has value to show the doing or not doing of the act is the inward emotion, passion, feeling, of the appropriate sort". It is clear that he saw emotion as a deeply psychological concept.

9. Framework of Motive Evidence

The general system of reasoning to or from a motive on either model has the structure of a five-tuple $\{M, F, A, S, D\}$. M is a motive. F is a set of statements representing the facts in a case. A is a set of argumentation schemes, most notably including the schemes for practical reasoning and abductive reasoning, used to draw conclusions. S is a set of story schemes. D is a set of dialogues of different types, including deliberation dialogue and persuasion dialogue. A dialogue has rules (protocols), moves, speech acts and commitment sets, of the kinds illustrated in Walton and Krabbe (1995). The system is applied in a standard case as follows (Walton, 2006b). The primary agent has carried out some actions and made some statements that are known by the secondary agent. The secondary agent can observe or hear them directly or can come to know them through sources such as testimony. Using F and S, the secondary agent uses story schemes to produce explanations. The inferred conclusions are fitted into S using the dialogue D to test an explanation. The secondary agent uses IBE to construct plausible hypotheses about the desires and actions of that other agent, and then judges which explanation is the most plausible, according to the given data (Walton, 2006b). The secondary agent can reconstruct the plans and actions of the primary agent because both agents have a grasp of familiar kinds of story scripts that are common in everyday experience.

What makes the explanation comprehensible is that both participants are agents. The primary agent deliberates on how to act in a given situation, facing a problem. The secondary agent knows facts describing what the primary agent did and understands how that agent was trying to solve a problem. The explanation is based on IBE story schemes. The process of simulative and abductive reasoning used to draw inferences from or to a motive takes place at two levels. At the primary level, the primary agent is engaged in deliberation on how to act by choosing among alternative courses of action in a given situation. At the secondary level, the secondary agent is engaged in asking questions about what the primary agent's reasons were for his or her actions. The shift from the one level to the other is possible because both participants are agents familiar with stereotypical stories who can seek out a solution to a practical problem by the same process of reasoning.

What the primary agent actually did and said is described. The secondary agent then takes the set of facts F and asks various kinds of questions about F. Under the list of six critical questions matching the scheme for practical reasoning given in Section 1, subquestions can also be asked (Walton, 1990), producing a much more extensive list. It can be conjectured here that the questions given by Bex, Bench-Capon and Atkinson (2009, 84) are particularly useful, because they can serve as a general set of critical questions when applying either the commitment-based scheme or the BDI scheme to evidential reasoning about motives.

CQ_1: Are there alternative ways of explaining the current circumstances S?

CQ_2: Assuming the explanation, is there something that takes away the motivation?

CQ_3: Assuming the explanation, is there another motivation that is a deterrent for doing the action?

CQ_4: Can the current explanation be induced by some other motivation?

CQ_5: Assuming the previous circumstances R, was one of the participants in the joint action trying to reach a different state?

CQ_6: Are the current circumstances true?

CQ_7: Could the action have had the stated preconditions?

CQ_8: Were the previous circumstances the same as the current circumstances?

CQ_9: Could the explanation for the current state provide the motivation?

CQ_{10}: Assuming the previous circumstances, would the action have the stated consequences?

CQ_{11}: Assuming the previous circumstances, would the action have any consequences?

CQ_{12}: Are the current circumstances S possible?

CQ_{13}: Is the joint action possible?

CQ_{14}: Are the previous circumstances R possible?

CQ_{15}: Is the motivation indeed a legitimate motivation?

To try to get answers to these questions, the secondary agent uses IBE to reason backward as a practical inference. To carry out this abductive task of questioning and inference, the secondary agent must try to reconstruct the deliberations of the primary agent, as they presumably took place at the primary level. By using IBE story schemes the secondary agent can infer that the primary agent had a motive in mind when he acted the way he did. Of course, this conclusion is only one explanation, based on IBE from the given evidence. The secondary agent cannot know for sure what the primary agent had in mind. But the conclusion drawn can be drawn as a more or less plausible explanation of the facts.

10. Conclusions and Further Research Directions

Leonard (2001, 449) has a very interesting account of how evidence concerning intention in law can be derived from motives evidence. To illustrate his point, let's go back to the same murder case we earlier considered, where D admitted killing V but claimed it was an accident. The evidence of D's threat to reveal V's auto theft, however, would function as evidence to the contrary. By offering an inference to the best explanation showing that D had a motive for killing V, an inference can be drawn that D acted intentionally and that this action was not merely accidental. Leonard structures this argument (449) as an inference of the following kind.

Evidence: D stole a car, V was aware of the fact, and V threatened to inform the police.

Inference: D had a motive to prevent V from revealing the theft to the police.

Conclusion: D purportedly killed V to prevent V from revealing the theft of the police.

Here we have an inference based on factual evidence and motive evidence leading to a conclusion about an agent's intention. This form of argument can be called argument from motive to intention. The postulation of a motive as premise is not by itself sufficient to prove intention, but motive along with other factual evidence can be used as an argument to bring forward evidence that might be sufficient to prove the existence of a motive.

Figure 6.9 shows the ultimate conclusion, attributing an intention to the defendant, on the left, supported by an argument from motive to intention (MtoI). On the right, other evidence uses abductive reasoning (IBE) to support the conclusion that the defendant had a motive. Typically, the motive to intention argument would have more than one premise. Other facts of the case would be needed to provide additional supporting evidence, strengthening the argument from motive to intention, which may

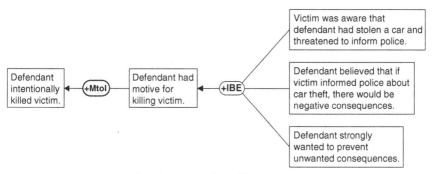

FIGURE 6.9 Argument from Motive to Intention

be a not very strong hypothesis to begin with. This example nicely illustrates the nature of the connection between motive evidence and intention.

The conclusion of this chapter is that inference to the best explanation based on the BDI model is the form of evidence-based argumentation for reasoning abductively in legal argumentation from an agent's actions to a hypothesis about the motive that may have led to the action. When practical reasoning is used for this purpose, its structure can be defined using the following argumentation scheme, the BDI scheme. The pronoun 'I' refers to an autonomous BDI agent capable of carrying out action based on its knowledge of a given situation, as well as its intentions and desires.

Major Premise: I have a desire *D*.

Minor Premise: Carrying out this action *A* is a means to realize *D*.

Conclusion: Therefore, I carry out action *A*.

In this model, an emotion is a particular desire that is singled out in the abductive reconstruction of an agent's action as that special desire or want that was supposedly the mainspring of the action. In other words, a motive is a particularly lively desire or want that causes an agent to carry out particular action to fulfill that desire.

The difference between desires and goals is that goals can be set or withdrawn deliberately, whereas desires are subject to control in a less direct way. A goal can also provide evidence for a motive in cases of deliberate planning to execute an action, but in many of the most typical cases in criminal law the action may be impulsive, not characterized by planning and careful comparison of the alternatives before taking action. In such cases, the motive can provide evidence that the person who is taken to have that motive has carried out particular action. As such, motives sometimes conform better to the practical reasoning structure of the BDI model, because a motive can be so lively just before the carrying out of the action that it often explains inference to a conclusion to act. The use of the BDI model is connected with heuristics, fast ways of arriving at a conclusion that can often be reasonable but that in some instances are associated with fallacies, biases and errors (Walton and Gordon, 2009).

The general system outlined above can accommodate the ambiguity noted by Wigmore. In one meaning of the term noted by Wigmore, the plaintiff might contend that the defendant's motive for burning down his house is the fact of the plaintiff's prior prosecution of a lawsuit against him. But as also noted by Wigmore, someone might also properly refer to the defendant's hostile and vindictive emotion arising from the lawsuit as the motive. The commitment model can be applied to a description of a motive of the former type, while the BDI model can be applied to the latter type.

The two models, the commitment model and the BDI model, can be applied in tandem. The commitment model displays the structure of practical reasoning in planning and rational deliberation where goals have been formulated and proposals for action are put forward, and where these actions are recommended as means to achieve the goals. The BDI model can be applied in cases such as those in the law, where a lively emotion can function as an explanation for why an agent presumably carried out some action, and then secondarily it can function as a means of using circumstantial evidence to infer backward to the presumed motive on which an action was based. Wigmore clearly thought that the view of motive as emotional desire that drives an agent to action is the deeper theoretical approach to evidential reasoning about motives in law.

7

The Carneades Model of Scientific
Discovery and Inquiry

In this chapter, the Carneades Argumentation System is used to model an example of the progress of scientific inquiry starting from a discovery phase. This new procedural bounded rationality model of scientific inquiry is used to show how a hypothesis can justifiably be accepted based on a process of marshaling and testing of evidence pro and contra, once it has been supported strongly enough by this evidential procedure to meet a standard of proof appropriate for the inquiry. Both discovery of a hypothesis and its later proof is seen as part of an orderly rule-governed procedure, modeled by a formal dialectical structure in which evidence is collected, tested and measured against standards of proof, then used to draw a justified conclusion. This context of argumentation was called the inquiry dialogue in Chapter 1.

The model supports an approach to scientific inquiry that could be classified as pragmatic, in that it varies with the standards of proof appropriate for kinds of inquiry in a field of knowledge and with criteria for it to be considered to be evidence. It is based on the theories of inquiry of Peirce (1931; 1984) and Popper (1963; 1972). According to the Carneades model of inquiry (Gordon, Prakken and Walton, 2007; Gordon, 2010), a group of interacting agents is collecting evidence as part of a search for the truth of a matter that they are collaboratively investigating. As they go along during the search process, they verify or falsify hypotheses by testing them using the data they have collected so far, at the same time as they are engaged in the process of collecting new data. As the search for knowledge continues, some hypotheses become better supported by the evidence, but at the same time, some of the hypotheses previously accepted have to be given up, because they are falsified by the new data that are streaming in.

In Section 1, formal models of dialogue of the kind used to analyze and evaluate argumentation are explained, in certain respects more fully than in Chapter 1, and some of their properties are stated. It is explained briefly how the Carneades computational model of argumentation is built around a formal dialogue system that incorporates argumentation schemes representing types of arguments used in evidential inquiries. In Section 2, a case study of scientific discovery is presented and the argumentation scheme for abductive reasoning is introduced. The case study has to do with causal reasoning, and in Section 3 an argumentation scheme is introduced, called argument from correlation to causation, that is important for understanding the case. In Section 4, a model of the type of dialogue called the inquiry is set out and its properties are explained. It is shown how the traditional model of inquiry is limited when it comes to modeling real instances of scientific investigation, and why a different model, based on the philosophies of science of Peirce and Popper (outlined in Section 5), needs to be used. The Carneades view of inquiry dialogue is presented in Section 6, and a framework called discovery dialogue is introduced in Section 7. The problem then posed is how discovery dialogue, where new hypotheses are accepted provisionally, based on very little evidence that has been marshaled so far, can shift to a later phase of inquiry dialogue in which the hypothesis is proved by showing that it meets an appropriate standard of proof. In Section 8, the contrast between discovery and the later proof phase of the scientific investigation is modeled by introducing the concept of a dialectical shift. In Section 9, it is shown how Carneades models burdens of proof shifts of a kind that are essential to understanding inquiry dialogue. The conclusions are in Section 10.

1. The Carneades Argumentation System

Argumentation can be defined as the technique for evaluating arguments that considers different arguments for and against some conclusion and how they support or rebut each other to determine which side has the stronger argument. As noted on the Web page for the Sixth International Workshop on Argumentation in Multi-Agent Systems[1] over the last few years, argumentation has been gaining increasing importance in artificial intelligence, especially in multiagent systems, where it has been used as a way to structure interaction, for example, on the Internet, that involves the giving and receiving of reasons. Two of the most useful tools for analyzing sophisticated forms of interaction among rational agents are argumentation schemes and normative models of dialogue that represent the context in which an argument was put forward for some purpose, for example, to resolve the conflict of opinions or prove a scientific hypothesis. The study of argumentation schemes, forms of argument that capture stereotypical patterns of human reasoning, is at the core of argumentation research. Schemes have been put forward as a helpful way of characterizing structures of human reasoning that have proved troublesome to view deductively or by

---

[1] http://homepages.inf.ed.ac.uk/irahwan/argmas/argmas09.

inductive models of reasoning based on Bayesian rules of inference using numerical calculations of probabilities. The other central tool of the argumentation method is the use of formal models of dialogue to represent the setting in which an argument or explanation was used for some purpose.

Dialogue models of argumentation of the kind developed in Walton and Krabbe (1995) are now proving their worth as tools useful for solving many problems in argumentation studies, artificial intelligence and multiagent systems. Many formal dialogue systems have been built (Bench-Capon, 2003; Prakken, 2005; 2006), and through their applications (Verheij, 2003), we are getting a much better idea of the general requirements for such systems and how to build them. Reed (2006) has provided a dialogue system specification that enables anyone to construct a formal dialogue model of argumentation by specifying its components and how they are combined (Reed, 2006, 26). Walton and Krabbe (1995) identified six primary types of dialogue: information-seeking dialogue, inquiry, deliberation, persuasion dialogue, negotiation and eristic (quarrelsome) dialogue. Argumentation has made solid contributions in computing to structuring multiagent dialogues that include legal disputes, business negotiations, labor disputes, team formation, deliberative democracy, risk analysis, scheduling and logistics.

Argumentation has also been applied to scientific discovery (McBurney and Parsons, 2001b) and inquiry (Black and Hunter, 2007). While building formal models of dialogue representing different contexts of argumentation, Walton and Krabbe (1995) also studied dialectical shifts, transitions from one type of dialogue to another. In this chapter these models are applied to analyzing the argumentation in a case of scientific discovery.

Formal dialogues are purely abstract normative structures that may be meant to model real dialogues, such as parliamentary debates or phone conversations, but they are only abstract structures that represent models of how the two parties should act and react if they are being "rational". The definition of rationality depends on what type of dialogue they are supposedly engaging in and what the rules for that type of dialogue are.

A dialogue is defined as (1) a set of participants, two, in the simplest case (2) taking part in a dialogue procedure (3) of a certain type. The two parties can be called the proponent and the respondent, or even more neutrally, Black and White. A dialogue is essentially a structure in which argumentation (including related types of activities such as questioning or explanation) take place in a rule-governed, orderly way as a transaction between the two participants. A dialogue is defined in the Carneades model as an ordered three-tuple $\langle O, A, C \rangle$ where O is the opening stage, A is the argumentation stage, and C is the closing stage (Gordon and Walton, 2009, 5). Dialogue rules define which types of moves are allowed (Walton and Krabbe, 1995).

At the opening stage, the participants agree to take part in some type of dialogue that has a collective goal. Each party has an individual goal and the dialogue itself has a collective goal. The type of dialogue is determined by its initial situation, the collective goal of the dialogue shared by both participants, and each individual participant's goal. The initial situation is framed at the opening stage, and the dialogue always then moves through the opening stage toward the closing stage. The six basic types of dialogue so far defined are persuasion dialogue, inquiry, negotiation dialogue, information-seeking dialogue, deliberation and eristic dialogue (Walton and Krabbe, 1995). The type of dialogue that has been studied most intensively so far is the persuasion dialogue.

During the opening stage (1) the collective goal is set in place, depending on the type of dialogue, (2) a database of statements is set up representing the shared knowledge of participants and the evidence collected up that point, and (3) the requirements for achieving the goal are determined. For example, in an inquiry dialogue, the collective goal is to prove a designated statement, or if it cannot be proved by the evidence collected, to prove. that it cannot be proved. The database of shared knowledge could be any database of statements accepted by the participants as representing knowledge. For example it could be the contents of an encyclopedia or the contents of authoritative textbooks. It also could represent a set of statements that can be taken to be common knowledge, meaning that these statements are not subject to dispute and would not be challenged by either party during the argumentation stage of the dialogue. The requirements for achieving the goal typically involve a burden of proof distributed on both parties. A burden of proof and a standard of proof that is part of the burden will be set up and allocated to each party. For example, the best argument standard is the rule that whichever side put forward the stronger argument wins the dialogue. While this standard is typical of deliberation dialogue, a much higher standard, like that for beyond reasonable doubt, is typical for inquiry dialogue. During the closing stage it is determined, according to the burden rule set at the opening stage, which party has won.

The argumentation stage A of a dialogue in the Carneades Argumentation System, as it is now called, is made up of a sequence of moves, where each move M is an ordered pair $\langle SpA, Con \rangle$, where A is the content of the move and SpA is a speech act representing the type of move whereby A was put forward in D. For example, the content of a move might be a statement, like 'Snow is white', and the speech act might be that of assertion. As each party makes a move, statements are inserted into or retracted from his or her commitment store. Commitment rules determine when and how insertions and retractions take place (Walton and Krabbe, 1995). Still other dialogue rules determine which side has achieved its individual goal (won or lost), once the closing stage of the dialogue is reached.

Four types of rules govern a dialogue: locution rules, dialogue rules, commitment rules and closing rules (Walton and Krabbe, 1995). The locution rules determine which kinds of moves can be made. The dialogue rules define the ordering of how participants take turns making moves and, in particular, which type of move is allowed by one party as a response to the previous move made by the other party. A commitment set is a database that keeps track of the commitments of each party that are incurred at each move he or she makes in the dialogue. The commitment rules, as noted above, determine which statements are inserted into or retracted from commitment sets of each party. Closing rules determine when the argumentation stage is closed off and whether one side or the other has achieved his or her goal in light of the argumentation put forward by both sides during the argumentation stage.

In Walton (2011b), a Carneades model of inquiry is presented in which a proposition is proven to be knowledge if it is accepted and supported strongly enough by the evidence to meet an appropriate standard of proof. In this model, which represents an evidentialist theory of knowledge, a proposition p does not have to be true to be included in knowledge. What constitutes proof varies with the standards of proof appropriate for kinds of inquiry in a given field of knowledge and with criteria for it to properly be considered as evidence. According to this model of scientific inquiry, a group of agents is trying to prove or disprove a hypothesis, and they accept a proof standard that enables the investigation to determine whether or not a proposition is proved.

According to the defeasible evidentialist notion of knowledge postulated in Walton (2011b), a proposition rightly said to be known to be true at a given point in the investigation could later turn out to be proved to be false. Contrary to the dominant view that only a true proposition can qualify as knowledge, on this evidentialist view, knowledge is not defined as justified true belief. On this dialectical model of inquiry, whether a proposition is rightly said to be knowledge or not depends on its rational acceptance, given the evidence for it, as balanced against the evidence against it, at the closing stage of the investigation. Knowledge is based on three factors (Walton, 2011b, 139): (1) the evidence collected at a given point in the investigation, (2) the kinds of arguments that can properly be used to justify a claim in that type of investigation and (3) the standard of proof set for knowledge in that particular type of investigation.

The analysis of argumentation in scientific discovery and proof that will be given below is based on the structure of a shift from a discovery dialogue to an inquiry dialogue. A discovery dialogue is characterized by an initial situation in which there are some data and where tentative hypotheses are constructed and tested that offer competing explanations of the data. For the statement asserted by the hypothesis to be proved, however, the argumentation has to continue to a subsequent phase. The inquiry phase has the goal of proving such a hypothesis to a higher proof standard so that it can be accepted as scientific knowledge, disproving it, or proving that it cannot be proved or disproved. It will be shown how burdens and standards of proof are important not only for defining both types of dialogue, but also for helping to identify the shift from the one type of dialogue to the other in cases of scientific discovery and proof.

2. The Cure for Anemia Case

In a classic case of scientific discovery, researchers persisted with methods that appeared very strange at the time (Jacovino, 1998). Pernicious anemia, a disease characterized by dangerously low counts of red blood cells, killed 6,000 people every year until 1923 when George R. Minot and William P. Murphy set out to find a cure. They were aware of some previous research by George Whipple, who found that the best treatment for anemic dogs was to feed them liver. When they tested this same treatment on their human patients, they observed an increase in the red blood cell counts of these patients. After testing the treatment on more patients, they found that it worked. For this research, Minot, Murphy and Whipple shared the Nobel Prize in 1934.

Once the cure was found, it led to further experiments and eventually to a scientific explanation of why it worked. Although it had been found that the cure worked, the question of what the active ingredient was in the liver that caused the increase in red blood cells, and the subsequent return to health of the patients, had not yet been answered. This question provoked further research that eventually led to an answer during a second stage. William Castle noted that people with their stomachs removed because of cancer often died because of anemia and was led by a series of experiments to formulate the hypothesis that something in the stomach was related to the disease. As an experiment, he ate red meat, forced himself to vomit, and then had patients eat his regurgitated stomach contents (Jacovino, 1998, 3). This revolting experiment led to scientific inquiry that proved that the substance necessary for red blood cell formation (found abundantly in liver) was vitamin B12. The substance that was the causal factor in liver remained unknown until 1948, when it was isolated by two chemists, Karl A. Folkers and Alexander R. Todd. They named the substance "vitamin B12", but it was not completely purified or identified until the 1950s.

The sequence of argumentation in the example takes the form typical of many examples of reasoning to a hypothesis in the experimental sciences. The first point in the sequence was the finding that feeding raw liver to animals or humans worked as a cure for anemia. An increase in the red blood cell counts of patients treated in this way was found. The question posed was this one: What active ingredient in the liver caused the increase in red blood cells? At this point in the investigation, a causal relationship had been hypothesized between anemia and deficiency of red blood cells. Although what had been found so far was a causal relationship that was

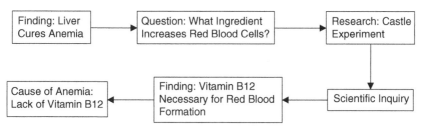

FIGURE 7.1 The Sequence of Argumentation in the Anemia Example

useful for clinical purposes, questions remained to be asked about the cause of anemia. The fuller sequence that the chain of argumentation took is shown in Figure 7.1.

The first two steps shown in Figure 7.1 led to the experiment by William Castle that showed that it was something in the stomach that was related to the disease. This experiment led to further scientific inquiry that gave a deeper explanation of the relationship between anemia and red blood cell formation by showing that the substance necessary for red blood cells was vitamin B12.

One form of argumentation that this case and other instances of scientific discovery are based on is abductive reasoning, or inference to the best explanation. The argumentation scheme representing this form of argument, along with its matching set of critical questions, was shown in Chapter 3, Section 7. An abductive inference (Josephson and Josephson, 1994, 14) has the following form, where H is a variable representing a hypothesis and D is a variable representing a given set of data or (presumed) facts.

D is a collection of data.
H explains D.
No other hypothesis can explain D as well as H does.
Therefore H is plausibly true.

On this account (Josephson and Josephson, 1994, 14), evaluating abductive reasoning in a given case depends on six factors.

1. How decisively H surpasses the alternatives
2. How good H is by itself, independently of considering the alternatives (we should be cautious about accepting a hypothesis, even if it is clearly the best one we have, if it is not sufficiently plausible in itself)
3. Judgments of the reliability of the data
4. How much confidence there is that all plausible explanations have been considered (how thorough was the search for alternative explanations)

5. Pragmatic considerations, including the costs of being wrong and the benefits of being right

6. How strong the need is to come to a conclusion at all, especially considering the possibility of seeking further evidence before deciding.

The conclusion to be inferred using these criteria is taken to be the best explanation of the given data. However, this form of reasoning is defeasible, and the conclusion may have to be withdrawn in the future, as new data comes to be taken into account in a case. A good example of abductive reasoning is scientific reasoning at the discovery stage, or early hypothesis construction stage of scientific research. It is a process of guesswork that proceeds by constructing hypotheses that would explain the given data.

In their classic chapter, Minot and Murphy (2001) related many observations made by themselves and others concerning the treatment of patients with pernicious anemia. These patients had been treated with many different kinds of diets. They described the effects of each type of diet and drew conclusions from these observations (342–349). They also examined results of experimental work concerning the effect of food on blood regeneration on the formation of hemoglobin. They reported the results of experiments of using ingestion of fresh red bone marrow. They used abductive reasoning repeatedly as they attempted to explain the outcomes of these experiments by drawing conclusions on how pernicious anemia affects factors such as red blood corpuscle count. They suggested hypotheses, but admitted that they were highly tentative in nature, commenting, "the spontaneous remissions of pernicious anemia and the bizarre course it often runs make it notoriously difficult to determine accurately the effect of any procedure on the disease" (350). They concluded that a diet rich in liver leads to a prompt rise in hemoglobin levels and often results in a marked subjective improvement in patients with pernicious anemia (351).

3. Causal Argumentation

It is important to realize that this case is an instance of the argumentation scheme for abductive reasoning, but another important aspect of it is that it is a causal argument designed not only to search for a scientific explanation, but also to prove a scientific conclusion. It also involves another argumentation scheme. In the argumentation scheme *CtoC* for arguments from correlation to causation (Walton, 2008a, 276), A and B are variables representing events or kinds of events.

There is a positive correlation between A and B.
Therefore A causes B.

The notion of positive correlation is relatively easy to define and measure. It means that wherever A has been observed, B has also, and the number of instances in which both events have been observed together can be counted. The causal relation is more difficult to define, but it means something roughly like the following. *A* is one of a set of conditions that are (when taken together) sufficient for the occurrence of *B*, and *A* is also a necessary condition for the occurrence of *B*. In addition, *A* is usually a condition of a kind that is subject to manipulation, so that we could, in principle, make *B* occur or prevent *B* from occurring if we can make *A* occur or prevent *A* from occurring.

Many instances of arguments that fit the scheme *CtoC* can be inherently reasonable, even if they are weak and subject to further investigation. In many instances, a positive correlation between two events can be some indication that there may be a causal connection between them. The problem is that many instances of *CtoC* can be misleading or even fallacious. For example, it may turn out on further investigation that some third factor accounts for the correlation, showing that the apparent causal relationship between *A* and *B* was misleading. This sort of error of jumping to a causal conclusion prematurely, merely on the basis of a correlation, has long been identified with the famous *post hoc, ergo propter hoc* fallacy.

The upshot is that *CtoC* should be seen as a defeasible form of argument that holds only tentatively and is subject to critical examination by the asking of critical questions. According to the account given in Walton (2008a, 277–278), there are seven critical questions matching the scheme *CtoC*.

1. Is there is a positive correlation between *A* and *B*?
2. Are there are a significant number of instances of the positive correlation between *A* and *B*?
3. Is there good evidence that the causal relationship goes from *A* to *B*, and not just from *B* to *A*?
4. Can it be ruled out that the correlation between *A* and *B* is accounted for by some third factor *C* (a common cause) that causes both *A* and *B*?
5. If there are intervening variables, then can it be shown that the causal relationship between *A* and *B* is indirect (mediated through other causes)?
6. If the correlation fails to hold outside a certain range of causes, then can the limits of this range be clearly indicated?
7. Can it be shown that the increase or change in *B* is not solely due to the way *B* is defined, the way entities are classified as belonging to the class of *B*s, or changing standards, over time, of the way *B*s are defined or classified?

Evaluating an argument from causation to correlation is best carried out in a dialogue format in which a defeasible argument is put forward by one side on an issue that, if supported by some evidence, shifts a burden of proof to the other side to respond by questioning the claim made. However, when an appropriate critical question is asked by the respondent, the burden of proof is thrown back onto the proponent's side. To meet this burden, and restore the acceptability of his or her argument, he or she must substantiate his or her causal argument by giving evidence that some other factor is not also at work, such as an intervening cause, a common cause or simply coincidence. As the proponent adequately answers each of the seven critical questions of the argument from correlation to causation, that initial suspicion can become more and more highly strengthened as an argument that fulfills its obligation in the discussion or inquiry.

What we should conclude then is that *CtoC* should always be seen as a double-edged sword. It can be a reasonable argument to accept a provisional hypothesis in some cases, but it is a form of argument that can be open to critical questions and be made stronger by answering these questions. We need to also be aware, however, that this form of argument can be highly misleading in some instances. For example, many traditional cures for common aliments and diseases did seem to their proponents to work, because there were many instances in which good outcomes were observed when the remedy was taken. However, when controlled scientific testing with placebos was undertaken, in many instances it was shown that the apparent benefit was merely an illusion. The *post hoc* fallacy is easy to commit in such cases. It is not as easy as it may appear to be to establish conclusively that there is a causal link between two states of affairs. To establish conclusively that A causes B, an investigator must arrive at a clear theoretical understanding of the mechanism whereby A is causally related to B. To prove a *CtoC* argument by an appropriate standard of proof, there needs to be an understanding of the underlying structural linkage between A and B as physical or causal processes. This means shifting the context of dialogue to that of a scientific inquiry.

4. Inquiry Dialogue

The goal of the inquiry is to prove that a statement designated at the opening stage as the *probandum* is true or false, or if neither of these findings can be proved, to prove that there is insufficient evidence to prove that the *probandum* is true or false (Walton, 1998, chapter 3). The aim of the inquiry is to draw conclusions only from premises that can be firmly accepted as true or false, to prevent the need in the future to have to go back and reopen the inquiry once it has been closed. The most important characteristic of the inquiry as a type of dialogue is the property of cumulativeness (Walton, 1998, 70). To say a dialogue is *cumulative* means that once a statement has been accepted as true at any point in the argumentation stage of the inquiry, that statement must remain true at every point in the inquiry through the argumentation stage until the closing stage is reached. Cumulativeness in an inquiry essentially means that once a statement has been accepted at some point, it can never be retracted at any next point.

The inquiry as a type of dialogue is somewhat similar to the type of reasoning that Aristotle called a demonstration. On his account (*Posterior Analytics*, 71b26), the premises of a demonstration are themselves indemonstrable, as the grounds of the conclusion, and must be better known than the conclusion and prior to it. He added (*Posterior Analytics*, 72b25) that circular argumentation is excluded from a demonstration. He argued that since demonstration must be based on premises prior to and better known than the conclusion to be proved, and since the same things cannot simultaneously be both prior and posterior to one another, circular demonstration is not possible (at least in the unqualified sense of the term 'demonstration'). As Irwin (1988, 123) explains, Aristotelian demonstration is ultimately based on first principles (axioms) that cannot themselves be proved. Trying to prove such principles by demonstrating them would be circular, and looping back in this way is incompatible with demonstration.

In contrast, persuasion dialogues, as well as deliberation dialogues and discovery dialogues, have to allow for retractions. It is part of the rationality of argumentation in a persuasion dialogue that if one party proves that the other party has accepted a statement that is demonstrably false, the other party has to immediately retract commitment to that statement. It does not follow that persuasion dialogue has to allow for retractions in all circumstances, but the default position is that it is presumed that retraction should generally be allowed, except in certain situations. In contrast, in the inquiry, the default position is to eliminate the possibility of retraction of commitments, except in certain situations. The normal pattern of managing commitment in the inquiry is to accept as conclusions only propositions that are firmly established, to prevent the danger that the argumentation might have to loop back into circular reasoning because a statement that was previously held to be firmly established has now been cast into doubt. An inquiry is always supposed to be a forward-moving sequence as it goes through the argumentation stage in which the foundations are firmly established. However, cumulativeness means not that retraction of commitments never occurs in argumentation in an instance of an inquiry, but only that the goal of the inquiry is to minimize or eliminate retractions insofar as possible, by setting a high standard of proof into place at the opening stage.

It was shown in Walton (1998, 72) how the structure of cumulative argumentation in the inquiry can be modeled by the semantics for intuitionistic logic presented by Kripke (1965). The model has a tree structure, where the nodes H_i are taken to represent what Kripke called "evidential situations" at a given point in an investigation at which some facts are known, but where more facts may come to be known at later points in the investigation. If a statement A is verified at a particular point H_i in an investigation, A is written above the node H_i in the tree. If A does not appear above a

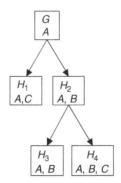

FIGURE 7.2 An Example of Cumulative Argumentation in a Kripke Model

node, it means that A has not yet been verified, at the point represented by the node.

Figure 7.2 (Walton, 1998, 73), represents advancing states of knowledge in an inquiry. In the example shown in the figure, the proposition A is verified at point G, but the proposition B is not. The inquiry could remain at G, or it could advance to H_1, where both A and C are verified. Alternatively, the inquiry could advance to H_2, where B is verified as well as A. Then the inquiry could then remain at H_2, or it could proceed to H_3 or H_4. At H_4, C is verified, in addition to A and B. At H_3, no new statements have been verified, but still the reader needs to notice that H_3 is a different evidential situation from H_2. At H_2, the possibility of verifying C remained open, whereas at H_3 new knowledge has now excluded the possibility of verifying C. The process of advancing states of knowledge in the Kripke model is cumulative, because once a statement appears at any node, it will appear at every accessible node as the tree exfoliates from that node.

It is a controversial issue among philosophers of science whether scientific argumentation should fit the model of the inquiry. The Kripke model of the inquiry certainly would not fit very well at the discovery stage; the case for applying it could be made to the stage where a scientist gets a presentation of experimental results ready for publication. When a researcher publishes scientific findings, in a field such as mathematics, physics or astronomy, great care is normally taken to show that the conclusion advanced is based on argumentation that employs the appropriate methods and standards of proof established in that field of science. The aim would appear to be removal or minimization of the necessity for later retraction of the claim. On the other hand, the normal requirement of falsification of hypotheses in scientific argumentation must leave open the possibility that retraction will be necessary in some cases. The aim of the inquiry, however, is that retraction will not be necessary.

As a case in point (Walton, 1998, 71), the retraction of the astronomer Andrew Lyne could be cited. He observed a pattern of periodically blinking stars and reached the conclusion that this observation could be explained only by the postulation of the existence of a previously undiscovered planet. After he re-examined his original analysis of the data, however, he concluded that it contained two small measurement errors that provided sufficient grounds for retracting his earlier hypothesis that the pulsing he had observed was caused by a new planet. It was an emotional moment when he had to admit his error at a scientific meeting. Initially, the audience felt pity for him, but at the end of his talk the audience applauded, evidently acknowledging the honesty and courage that must have motivated such an admission.

Cumulativeness appears to be such a strict model of argumentation that many equate it with the Enlightenment ideal of foundationalism of the kind attacked by Toulmin (1958). To represent any real instance of an inquiry, it is useful to explore inquiry dialogue systems that are not fully cumulative.

Rescher (1977, 101) takes a pragmatic approach to cognitive justification by adopting a dialectical methodology of inquiry. According to his description of the dialectical process of inquiry, there is a succession of states during which an initial thesis is tested, the evidence supporting it is put forward, and these two stages alternate with intervals where the arguments supporting the thesis are critically examined. Rescher portrays this dialectic of inquiry as a cyclical process in which a position is examined to find flaws in it by probing counterarguments, leading to an improved version of the initial position, and then once again this improved version is subjected to examination by probing counterarguments. This procedure continues with the result that the initial position is sequentially revised and continues to improve by taking into account the criticisms directed against it. These criticisms require more careful formulation of the position that includes qualifications that take the objections into account.

Adopting this pragmatic approach leads Rescher (1977, 110) to a disputational model of scientific inquiry in which the burden of proof rests with the proponent to bring forward evidence supporting his or her claim against the objections made by a respondent or group of critics. Experimentation plays a role in the examination process that probes into a theory at its weakest points to throw doubt on the claim or theory brought forward for acceptance. According to Rescher (1977, 121), this dialectical model of scientific inquiry accommodates both a confirmationist and a falsificationist view of science. Evidence supporting the claim needs to be evaluated by experimental testing and support of arguments. But at the same time, the arguments that an opponent brings forward to challenge or undermine the arguments made to support the proponent's claim constitute attempts at falsification of the hypothesis that is challenged.

McBurney and Parsons (2001b) have put forward a dialectical model of inquiry that allows for epistemic uncertainty and retraction of commitments by both parties in a dialogue during the process of scientific inquiry. Their model of inquiry (130) is built on the principle that the success of science as a form of inquiry rests on two basic normative principles. One is that every scientific explanation or hypothesis proposed by the scientific investigator is contestable by anyone. The other is that every argument, hypothesis or theory adopted by the scientific community is defeasible. From these principles, in their model it follows that the scientific conclusion, no matter how compelling, should always be treated as being tentative, that is, a hypothesis or theory that is being accepted but is later subject to retraction when a better hypothesis or theory is found.

Their model incorporates a number of features, including the following (McBurney and Parsons, 2001b 132–134): formulation of procedural rules, commitment to a proposition based on acceptance by a participant in a dialogue format, turn-taking in making moves in the dialogue, orderliness in dealing with issues one at a time, the following of an inquiry dialogue through a series of stages from an opening stage to the closing stage, permitting of both deductive and nondeductive forms of inference, rules that enable a participant to retract a previous commitment, allocation of burden of proof, and the permitting of direct mutual appeal to experience this evidence supporting a claim. On this basis they build a mathematical model of dialogue for scientific inquiry and illustrate how the procedure works through examples of scientific risk assessment of carcinogenic chemicals.

Black and Hunter (2007) have built a system of inquiry dialogues meant to be used in the medical domain to deal with the typical kind of situation in medical knowledge consisting of a database that is incomplete and inconsistent and operates under conditions of uncertainty. The kind of the inquiry dialogue they model is represented by a situation in which many different health care professionals are involved in the care of the patient and must cooperate by sharing their specialized knowledge in order to provide the best care for the patient. To provide a standard for soundness and completeness of this type of dialogue, Black and Hunter (2007, 2) compare the outcome of one of their actual dialogues with the outcome that would be arrived at by a single agent that has as its beliefs the union of the belief sets of both of the agents participating in the dialogue. Their current model assumes a form of cumulativeness in which an agent's belief set does not change during a dialogue, but they add that they would like to further explore inquiry dialogues to model the situation in which an agent has a reason for removing a belief from a belief it had asserted earlier in the dialogue (Black and Hunter, 2007, 6). To model real instances of argumentation inquiry dialogue, ways of relaxing the strict requirement of cumulativeness need to be considered.

5. Peirce and Popper on the Inquiry

Peirce's conception of the inquiry challenged the view that an inquiry cannot admit of retraction, opposing the view that a successful inquiry proves that the proposition proved in the inquiry has to be true in a manner implying that it cannot be subsequently doubted. He advocated the view of the inquiry based on epistemological fallibilism of a kind that admits of the susceptibility of scientific proof of a hypothesis to error. On his view, even the most careful scientific inquiry might produce an outcome that could change later, as new evidence comes in or as a new theory is accepted (Peirce, 1931, 2.75). Peirce wrote that many things are "substantially certain" (1.152), but that this is different from the kind of absolute certainty that implies truth. On his view truth is an aim of inquiry, and the motive of finding the truth is important for scientific research, but that truth can be arrived at beyond all doubt only during an inquiry that would take an infinite amount of time. Indeed, it was part of his view that the fixing belief by "tenacity" or "authority" is an important factor in blocking the way of inquiry by implying the claim that no further inquiry is necessary (Cooke, 2006, 34). To fix belief in this way is to claim that the proposition claimed is beyond question and no longer open to further attempts at evaluation and explanation.

Peirce took the unusual approach of understanding truth as part of the process of inquiry, instead of defining truth separately and using that definition to define the goal of an inquiry. He wrote (1986, 273), "The ideas of truth and falsehood, in their full development, appertain exclusively to the scientific method of settling opinion". He described the process of inquiry as one in which different participants in the inquiry set out with conflicting views that are ultimately carried by a force outside themselves to the same conclusion as a successful inquiry moves ahead. The whole process is described by him as being carried forward in a "predestinate" manner to this common opinion so that each separate agent in the process, although starting from a different opinion, is driven toward the common opinion (Tamminga, 2001, 20). Peirce wrote, "The opinion which is fated to be ultimately agreed to by all who investigate, is what we mean by the truth, and the object represented in this opinion is the real" (Peirce, 1986, 273).

Misak (1991) argued that Peirce's remarks on truth should not be interpreted as an attempt to define the concept of truth, but that it can be seen as a way of further explaining the concept of truth presented by the Tarski definition. Misak explained this linkage between inquiry and truth by means of two subjunctive conditionals (Tamminga, 2001, 21). The first one states that if the hypothesis is true, then if the inquiry relevant to the hypothesis were to be pursued as far as it can go, the hypothesis would be believed (Misak, 1991, 43). The second one states that if the inquiry relative to the hypothesis were to be pursued as far as it could go, and through this process the hypothesis would come to be believed, then the hypothesis is true (Misak, 1991, 46). These two conditionals show how truth can be understood as part of the process of inquiry by linking it to the common beliefs of the inquirers at different points in the sequence of inquiry.

For Popper the search for truth is also an important motive for scientific discovery, but like Peirce, he held that the best a successful scientific inquiry can do is to approximate the truth. On his view, a claim established by inquiry may properly be considered scientific knowledge even though it is falsifiable. Indeed, he held that all genuine scientific hypotheses are falsifiable. He claimed that scientific knowledge is objective, in the sense that it is based on an evidential procedure that moves toward truth as its goal and is independent of the knowing subject. Both Peirce and Popper viewed the inquiry as a procedure that uses evidence pro and con a hypothesis to move forward to tentative acceptance of a scientific theory or experimental finding. Thus both views of scientific discovery, inquiry and proof adopt a view of bounded procedural rationality.

For Popper the procedure of conjecture and refutation begins with the formulation of a problem P_1 and goes through a process of marshaling evidence to a theory TT, a hypothesis that represents the solution to the problem that started the process. He called next stage in the procedure "error elimination" (EE), a "severe critical examination" of the conjecture in a critical discussion that compares competing hypotheses (Popper, 1972, 164). The next step P_2 is the restatement of the problem situation that has emerged from the procedure of testing. On Popper's representation of the structure of the procedure (164), it has this form: $P_1 \rightarrow TT \rightarrow EE \rightarrow P_2$. The procedure carries on through successive refinements of the statement of the problem P_1, P_2, ..., P_n in a recursive manner. The procedure as a whole was seen by Popper (1972, 312) as a continuous movement of trial and error that proceeds by degrees of improvement and that can reach an outcome that is provisionally acceptable. Once a belief has been settled by the scientific community on the basis of inquiry, it can be accepted as true. However, it needs to be emphasized that on his theory, even though a theory or hypothesis can be accepted as part of science, it must always remain open to falsification or otherwise it does not count as genuine scientific knowledge.

6. The Carneades Model of Inquiry

Recent work in artificial intelligence and argumentation has formulated standards of proof that are designed to model legal argumentation (Gordon, Prakken and Walton, 2007), that can also in some instances be applied to other contexts, such as that of scientific argumentation. However, the legal standards are formulated in law in terms of how credible and convincing an argument is to the mind and to removing doubt. To formulate standards of proof in a more precise way that might be useful as applied to scientific argumentation, Gordon and Walton (2009) have defined an abstract formal model of argumentation as a theory and proof construction process for making justified decisions that can be applied to scientific as well as legal argumentation. The formal model is based on definitions of argument and dialogue consistent with the definitions of these terms presented in Section 2 above. In the model, four standards of proof are formulated, as recapitulated below. It is assumed that there can be more than one argument supporting a claim and that there can be pro arguments as well as con arguments with respect to a claim. It is also assumed that arguments can be comparatively weighed so that one argument is said to be stronger than another based on the evidence for and against each of the arguments.

- The scintilla of evidence standard is met iff there is one argument supporting the claim.
- The preponderance of the evidence standard is met iff the scintilla of evidence standard is met and the weight of evidence for the claim is greater than the weight against it.
- The clear and convincing evidence standard is met iff the preponderance of the evidence standard is met and the weight of the pro arguments exceeds that of the con arguments by some specified threshold.
- The beyond reasonable doubt standard is met iff the clear and convincing evidence standard is met and the weight of the con arguments is below some specified threshold.

These ways of formulating standards of proof are inspired by legal reasoning, and different standards can be applied in scientific reasoning depending on the context of dialogue in the case.

The problem is to know how to use Carneades to represent the notion attributed to Peirce that at some point a scientific finding can be accepted as substantially certain, so that at that point in a successful inquiry we can say that for all practical intents and purposes this proposition can be accepted as true. As we noted, Popper has a similar idea that once a belief has been settled by the scientific community through inquiry, it can be accepted as true. The best way of modeling these views using Carneades is to view a proposition in an inquiry, once it has reached this settled state of being substantially certain, as a belief that is fixed when an inquiry dialogue is temporarily closed, even though it can be subject to reopening at some future point.

Peirce had interesting ideas about how belief is fixed that can usefully be brought to bear in solving this problem. Walton (2010a) provided a model of belief in a formal dialogue system built on Peirce's description of the fixation of belief. This model can help to explain how truth can be defined in an inquiry in the Peircean manner by linking it to how a belief comes to be fixed in a scientific inquiry and is shown by Misak (1991) and Tamminga (2001). This dialogue model postulates ten characteristics of belief.

1. It is opposed to doubt, an uneasy and dissatisfied state.
2. It is a settled state, a "calm and satisfactory state".
3. It is a state we do not wish to change.

4. It is something we cling tenaciously to.
5. We cling to believing what we believe.
6. It can be firmly fixed, as with fanatical believers.
7. It is an indication of a habit.
8. It is a matter of degree.
9. It puts us into a condition so we act in a certain way in the future.
10. It both guides our desires and shapes our actions.

To summarize, a belief is defined in the dialogue model as a proposition held by an agent that (1) is not easily changed (stable), (2) is a matter of degree (held more or less weakly or strongly), (3) guides the goals and actions of the agent and (4) is habitually or tenaciously held in a manner that indicates a strong commitment to defend it. On the theory of Walton (2010a), beliefs are derived abductively by one participant in a dialogue from the commitment set of the other participant using evidence collected so far in the dialogue. A group of scientific investigators may reach a point in an inquiry where they firmly fix their belief that a particular proposition is true because it can be proved to be true to an appropriate standard of proof on all the evidence marshaled during the inquiry.

As the views of Popper and Peirce on the inquiry are represented in the Carneades model, the procedure used to judge a hypothesis as acceptable has to be based on evidence that is tested. The inferential procedure that takes us from the evidence pro and contra a knowledge claim for a proposition p to the conclusion that p is acceptable (or not) is shown in Figure 7.3. As shown in Figure 7.3, a hypothesis is subject to testing by marshaling and evaluating evidence for and against it. As the process of accumulating and testing evidence moves forward, objections and refutations may lead the inquiry to reject a proposition that was formerly accepted. The need for us to occasionally reject propositions that are shown to be false by new evidence is required by the views of inquiry of Peirce and Popper that see it as an instance of use of argumentation in a framework of bounded procedural rationality.

According to the sequence of argumentation shown in Figure 7.3, at the opening stage what counts as evidence must be specified, according to the field and nature of the inquiry. At this stage the standard of proof is set. During the argumentation stage, evidence for and against the proposition is brought forward and evaluated. As determined by the availability of time and resources, the sequence of argumentation must eventually be closed off.

During the closing stage, as shown along the bottom of Figure 7.3, the standard of proof is applied to determine whether the proposition that was the subject of the inquiry can be said to have been proved. One type of outcome may be that there was insufficient evidence to prove the proposition

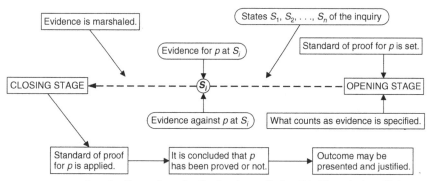

FIGURE 7.3 Inquiry Procedure for Evaluating a Defeasible Knowledge Claim

according to the required standard, and therefore it must be concluded that the proposition is unproven. It can be argued that this pragmatic model of knowledge is more useful than the prevailing epistemological one, because it arguably shows better how the external standard should be applied to reality in an inquiry.

It is not possible to set a single standard of proof for every scientific investigation. There is also the question of the conditions under which an inquiry should be reopened. From a fallibilist point of view, it is unrealistic to set a standard of beyond all doubt, and it is necessary even if one wants to set a very high standard, such as that of beyond reasonable doubt, to leave open the possibility that the inquiry can be reopened for further scientific investigation bringing in new evidence. This assumption is based on the defeasibility of scientific knowledge, which is in turn based on falsifiability as a criterion of genuine scientific knowledge.

On this view there is not the same sharp discontinuity between scientific discovery and proof found in earlier philosophies of science. The two phases of scientific inquiry are connected in a continuous process that moves toward proof and truth.

7. Discovery Dialogue

Discovery dialogue appears at first be the direct opposite of inquiry dialogue, because discovery dialogue is highly creative. It involves brainstorming and thinking up new hypotheses that still have not been tested and might well turn out to be false. Clearly, then, discovery dialogue requires quite a wide latitude for retracting previous commitments. Proving and disproving seems to be much more tightly controlled and restrictive in an inquiry dialogue than it could be in a discovery dialogue. In other ways, however, of all the six basic models of dialogue, discovery dialogue seems to be closest to the inquiry type, because in both types of dialogue participants are collaborating to ascertain the truth of some question at issue.

Discovery dialogue was first recognized as a distinct type of dialogue different from the any of the six basic types of dialogue by McBurney and Parsons (2001a). On their account (, 2001a, 417)), discovery dialogue and inquiry dialogue are distinctly different in a fundamental way. In an inquiry dialogue, the proposition that is to be proved true is designated prior to the course of the argumentation in the dialogue, whereas in a discovery dialogue the question whose truth is to be determined emerges only during the course of the dialogue itself. According to their model of discovery dialogue, participants began by discussing the purpose of the dialogue, and then during the later stages they use data items, inference mechanisms and consequences to present arguments to each other. Two other tools they use are called criteria and tests. Criteria, like novelty, importance, cost, benefits and so forth, are used to compare one data item or consequence with another. The test is a procedure to ascertain the truth or falsity of some proposition, generally undertaken outside the discovery dialogue.

The discovery dialogue moves through ten stages (McBurney and Parsons, 2001a, 419), called open dialogue, discuss purpose, share knowledge, discuss mechanisms, infer consequences, discuss criteria, assess consequences, discuss tests, propose conclusions and close dialogue. The names for these stages give the reader some idea of what happens at each stage as the dialogue proceeds by having the participants open the discussion and discuss the purpose of the dialogue. They go on to share knowledge by presenting data items to each other, to discuss the mechanisms to be used, such as the rules of inference, and to build arguments by inferring consequences from data items. They then discuss criteria for assessment of consequences presented, assess the consequences in light of the criteria previously presented, discuss the need for undertaking tests of proposed consequences, pose one or more conclusions for possible acceptance and close the dialogue. The stages of the discovery dialogue may be undertaken in any order and may be repeated (6). Agreement is not necessary in a discovery dialogue, unless the participants want to have it.

McBurney and Parsons also present a formal system for discovery dialogue in which its basic components are defined. A wide range of speech acts (permitted locutions) that constitute moves in a discovery dialogue include the following: propose, assert, query, show argument, assess, recommend, accept and retract. There is a commitment store that exists for each participant in the dialogue containing only the propositions that the participant has publicly accepted. All commitments of any participant can be viewed by all participants. They intend their model to be applicable to the problem of identifying risks and opportunities in a situation where knowledge is not shared by multiple agents. However, they do not (at least directly) consider how their model might be applied to scientific discovery in the kind of case where a new scientific finding was first discovered. It would seem, however, that their model could be applied to such cases.

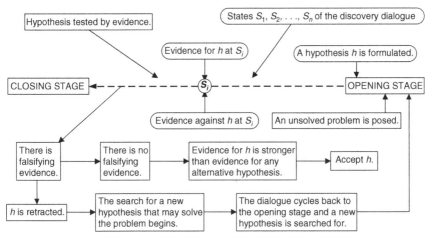

FIGURE 7.4 Overview of the Dialogue Model for Scientific Discovery

In a scientific discovery dialogue, an initial problem is posed or given that needs to be solved or explained by scientific methods. At this opening stage, discussing the purpose of the dialogue is appropriate. Although the purpose is one of discovery of something, it could be a problem in theoretical science that needs to be solved or a discovery of how to do something practical, like find a cure for a disease. The eight stages of McBurney and Parsons, before the closing stage, fit into the argumentation stage.

It might seem that discussing the mechanisms and criteria to be used should take place at the opening stage, but it seems even better not to lock these mechanisms and criteria into place at the opening stage, and then to make them binding through the argumentation stage. Discovery dialogue needs to be more open and flexible, so that the stages in the argumentation part can be undertaken in any order and repeated, as proposed in the model of McBurney and Parsons. For these reasons, discovery dialogue needs to be seen as quite different from inquiry in an important respect. In an inquiry, the burden of proof, including the standard of proof, needs to be set at the opening stage. Once it is set, it applies to the entire sequence of argumentation put forward by both sides during the argumentation stage. The burden of proof and the standards of proof are then used at the closing stage to determine whether or not the inquiry has been successful. An overview of the main stages is presented in Figure 7.4.

Discovery dialogue is quite different from inquiry dialogue in this regard. The procedure in a discovery dialogue looks similar in basic outline to the structure of an inquiry, because it is basically one of continual testing against the pro and contra evidence as this evidence comes into the dialogue. But the way the argumentation is evaluated, based on this evidence, is quite different. In a discovery dialogue, as shown in Figure 7.4, the hypothesis is rejected as soon as any falsifying evidence comes in. When this happens, the procedure cycles back to the opening stage where a new hypothesis is formulated, and then it follows the same procedure of being subjected to testing by the evidence found. However, if no falsifying evidence is found, then the hypothesis can be tentatively accepted, provided there is no alternative hypothesis that provides a better explanation of the same data.

The burden of proof at the closing stage in a discovery dialogue is therefore quite different from the way burden of proof works in an inquiry dialogue. In an inquiry, the aim is to avoid retraction insofar as this is possible, and therefore a high standard of proof needs to be set. The conclusion accepted in a discovery dialogue is much more conjectural in nature. A hypothesis may be tentatively accepted even though the evidence supporting it is fairly slight, provided there is no evidence against it and there is no competing hypothesis that is more strongly supported by the evidence. The reason for this light burden of proof is that the aim of a discovery dialogue is not to prove something to a high standard of proof, but only to provide a hypothesis that enables an investigation to move forward by collecting and assessing more evidence that will be useful in moving toward an inquiry dialogue, once the evidence that has been amassed is sufficient to justify opening an inquiry.

8. Dialectical Shifts

The longstanding problem with modeling scientific reasoning in philosophy of science has been the sharp contrast between (1) the Enlightenment model taken over by the positivists holding that scientific proof fits a foundational structure, and (2) a less rigid model in which retractions are allowed. The answer appears to be that we need both models. But if we try to combine these models, there is a problem. How there can be a transition in the sequence of reasoning in a scientific investigation from the one model of argumentation to the other? In this section it is shown how the notion of a dialectical shift can be used to represent the transition.

A dialectical shift is said to occur in cases where, during a sequence of argumentation, the participants begin to engage in a type of dialogue different from the one in which they were initially engaged (Walton and Krabbe, 1995). In the following classic case often cited as an example, two agents are engaged in deliberation dialogue on how to hang a picture. Engaging in practical reasoning, they come to the conclusion they need a hammer and a nail, because they have figured out that the best way to hang the picture is on a nail, and the best way to put a nail in the wall is by means of a hammer. One agent knows where a hammer can be found, and the other has a pretty good idea of where to get a nail. At that point, the two begin to negotiate about who will get the hammer and who will go in search of a nail. In this kind of case, we say that the one dialogue is embedded in the other (Walton and Krabbe, 1995), meaning that the second dialogue fits into the first and helps it along toward achieving its collective goal. In this instance, the shift to the negotiation dialogue is helpful in moving the deliberation dialogue along toward its goal of deciding the best way to hang the picture. For, after all, if somebody has to get the hammer and nail, and they cannot find anyone who is willing to do these things, they will have to rethink their deliberation on how best to hang the picture. Maybe they will need to phone a handyman, for example. This example of an embedding is contrasted with an example of an illicit dialectical shift when the advent of the second type of dialogue interferes with the progress of the first. For example, let's consider a case in which a union–management negotiation deteriorates into an eristic dialogue in which each side bitterly attacks the other in an antagonistic manner. This kind of shift is not an embedding, because quarreling is not only unhelpful to the conduct of the negotiation but is antithetical to it, and may very well even block it altogether, by leading to a strike, for example.

There is a way to represent dialectical shifts in what is called a dialogue frame, defined as a four-tuple composed of (1) a type of dialogue, (2) a topic, (3) a pair of participants and (4) a sequence of utterances (Reed, 1998, 248). The type of dialogue, t, could be persuasion, negotiation, inquiry, deliberation or information-seeking. The topic, τ, is the issue to be resolved by the dialogue, set at the opening stage. The two participants in a dialogue are represented as x_i and y_i. A dialogue is a sequence of moves in which the participants take turns making utterances (locutions) of the permitted types. Each utterance made by a participant during the dialogue is numbered. Thus $u^i x_i \rightarrow y_i$ refers to the ith utterance in the dialogue. Each utterance is a pair in which the first element is a statement and the second element is a support for that statement. Using this notation, a dialogue frame F has the following form (248).

$$F = \langle\langle t, \Delta \rangle \in D, \tau \in \Delta, (u^0 x_0 \rightarrow y_0, ..., u^n x_n \rightarrow y_n)\rangle$$

To cite the picture-hanging example, one move could be marked as an utterance that is part of a deliberation dialogue, but the very next utterance in the sequence of argumentation could be marked as part of a negotiation dialogue. This markup would indicate the existence of a dialectical shift at that very point of transition.

In clear cases, as shown in examples in Walton (2007a, chapter 6), there is no difficulty in identifying the place in the sequence of argumentation where the shift occurred. In the formal model of such a dialogue, each shift will be marked. The dialogue sequence below, from Walton (2007a, 243), shows how such a dialogue could be represented in Reed's case of picture-hanging (Reed, 1998, 249).

1. White: *propose* (*deliberate* (*can, White, hang-picture*))
2. Black: *accept* (*deliberate* (*can, White, hang-picture*)
3. White: *propose* (*have* (*White, nail*) and (*can, White, hang-picture*))
4. Black: *accept* (*have* (*White, nail*) and (*can, White, hang-picture*))

The dialogue starts out as a deliberation in which each party makes proposals on how to fulfill a common goal, and each responds to the proposals put forward by the other side. At the first move, White proposes that he and Black hang a picture. At the second move, Black accepts the proposal. At the third move, White proposes that he has a nail and can hang the picture. At this third move, White and Black have shifted to a negotiation dialogue, assuming that both parties are aware that both a hammer and a nail are required to hang the picture, that only Black has the hammer and that only White has the nail. It is clear that the "proposal" amounts to an offer. The two of them are now making a deal. Black then accepts the proposal, and White will assume that, at the next move, Black will offer to supply the hammer. To identify the shift, one has to (1) reconstruct the dialogue sequence of argumentation, (2) know the respective goals of the two types of dialogue involved and (3) analyze the speech acts (type of moves) in the area where the shift occurred. One has to be able to identify when a proposal is merely a proposal to carry out a designated action, and when it is functioning as an offer that is an opening move for a negotiation. The model has to be applied to the text of discourse in the case.

An extensive case study is needed to examine details of particular cases of scientific argumentation, like the one chosen here as an example, and to use indicators to judge when the argumentation has shifted from a discovery dialogue to an inquiry dialogue. There could be many textual indicators of this kind of shift, but the central one we emphasize here as centrally important is that the standards of proof differ. In an inquiry, the standards of proof are set high, and fixed at the opening stage, in order to prevent a need for retraction during the argumentation stage. The aim is to find knowledge, and in order to qualify as knowledge, a claim must be proved to an appropriately high standard. In a discovery dialogue, the aim is to pick out what appears to be a plausible hypothesis from a set of possible alternative hypotheses in which some are more plausible than others, so that further tests and experiments can be carried out. The aim of such testing procedures will, in many instances, be to falsify the hypothesis. To choose a hypothesis, there needs to be some evidence to support it and no evidence against it (so far), but the assumption is that some such falsifying evidence may be found as further evidence comes in. Thus the proof standard can be variable and may not need to be all that high.

9. Burdens and Standards of Proof

To be able to identify when a dialectical shift from a discovery dialogue to an inquiry dialogue has occurred in a particular case, we first of all have to investigate how the one type of dialogue is different from the other. Most important, there are basic differences in how burden of proof, including the standard of proof, operates. In an inquiry dialogue the global burden of proof, which is operative during the whole argumentation stage, is set at the opening stage. In a discovery dialogue no global burden of proof is set at the opening stage that operates over both subsequent stages of the dialogue. McBurney and Parsons (2001a, 418) express this difference by writing that in inquiry dialogue, the participants "collaborate to ascertain the truth of some question", while in discovery dialogue, we want to discover something not previously known, and "the question whose truth is to be ascertained may only emerge in the course of the dialogue itself". This difference is highly significant, as it affects how each of the two types of dialogue is fundamentally structured.

In an inquiry dialogue, the global burden of proof is set at the opening stage and is then applied at the closing stage to determine whether or not the inquiry has been successful. This feature is comparable to a persuasion dialogue, where the burden of persuasion is set at the opening stage (Prakken and Sartor, 2007). At the opening stage of the inquiry dialogue, a particular statement has to be specified, so that the object of the inquiry as a whole is to prove or disprove this statement. In a persuasion dialogue, this burden of proof can be imposed on one side or imposed equally on both sides (Prakken and Sartor, 2006a). However, in an inquiry dialogue there can be no asymmetry between the sides. All participants collaborate together to bring forward evidence that can be amassed together to prove or disprove the statement at issue. Discovery dialogue is quite different in this respect. There is no statement set at the beginning in such a manner that the goal of the whole dialogue is to prove or disprove this statement. The basic reason has been made clear by McBurney and Parsons. What is to be discovered is not known at the opening stage of the discovery dialogue. The aim of the discovery dialogue is to try to find something, and until that thing is found, it is not known what it is, and hence it cannot be set as something to be proved or disproved at the opening stage as the goal of the dialogue.

Hence it follows that the burden of proof is quite different in these two types of dialogue. In inquiry dialogue, the burden of proof is set at the opening stage, governs the conduct of the argumentation through the whole argumentation stage and then is used at the closing stage to determine when the argumentation stage should end and whether the argumentation in it was successful or not in fulfilling the goal of the dialogue. In discovery dialogue, what is set at the opening stage is some puzzle or problem that needs to be solved. In scientific argumentation, hypotheses are constructed during the argumentation stage. They represent ways of solving the puzzle or problem. As the evidence comes in, by further observations and by testing the hypotheses, it may be shown that some are better supported by the evidence than others, and even that some are refuted by the evidence. One may even emerge as the "best" hypothesis, the one supported by the most evidence and least open to attacks that might refute it.

In scientific argumentation, the puzzle or problem that needs to be solved is generally some phenomenon or appearance that is not yet understood and needs to be explained. Typical of the argumentation used is the scheme for abductive reasoning, or inference to the best explanation (Magnani, 2001). Some problem or unexplained event is identified at the opening stage, and then during the argument stage a number of explanations of the event are considered. The evidence for one explanation is weighed against the evidence for a competing explanation or a set of competing explanations (Josephson and Josephson, 1994). In a successful discovery dialogue, sufficient evidence is brought forward to prove that one explanation is demonstrably better than the others. But when we say 'prove' here, we do not necessarily refer to deductive or inductive argumentation exclusively. The kind of argumentation that is typically used to support or attack abductive reasoning is inherently defeasible in nature. A tentative conclusion is arrived at as a basis for further discovery dialogue that may ultimately lead to an inquiry. Use of abductive reasoning in discovery dialogue typically does not result in a conclusive proof or disproof of the hypothesis being considered.

As noted above, the standard of proof in inquiry dialogue tends to be set very high, as compared with other types of dialogue such as deliberation or persuasion dialogue, because the aim of inquiry is to remove troublesome doubts. Thus the aim is to prove some designated statement, or prove that it cannot be proved, in such a manner that the necessity for later retraction is eliminated as much as possible. It is the way this burden of proof is set at the opening stage that is the leading fundamental characteristic of inquiry as a type of dialogue. However, a different kind of burden of proof, called burden of production, is operative during the argumentation stage (Prakken and Sartor, 2007). According to this burden, once a given argument is put forward, it has to be backed up with evidence if any participant in the inquiry expresses doubts about it. If enough evidence has been put forward to satisfy the questioner, and remove his or her doubts to a reasonable degree, the argument can be accepted as proved.

We can see how the Carneades model applies to the argumentation in an instance of causal reasoning by looking at Figure 7.5. As shown in the figure, there is a causal claim at issue, supported by a correlation between the two events that are claimed to be causally related. Critical questions

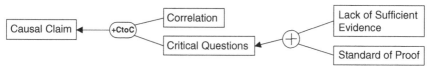

FIGURE 7.5 How Carneades Models Critical Questions

are modeled by Carneades as additional premises corresponding to the critical questions of an argumentation scheme. In the Carneades system, critical questions matching an argument are reformulated as three types of premises, called ordinary premises, assumptions and exceptions (Gordon and Walton, 2009). Assumptions are assumed to be acceptable unless called into question. Exceptions are modeled as premises that are assumed to be not acceptable and can undercut an argument if found to be acceptable. Ordinary premises are not assumed to be acceptable, but must be supported by further arguments in order to be judged acceptable. The correlation may be weak evidence to support the causal claim, but in the absence of support for the additional premises corresponding to critical questions, the evidence of correlation may be insufficient to support the causal claim. Whether or not the evidence is sufficient depends on the standard of proof, which in turn depends on the type of dialogue that is involved.

The problem with modeling critical questions so that they can be used with a standard argument visualization tool that uses a graph structure in which the nodes are statements is that there is no single way to represent the critical questions on the argument map. The solution to this problem shown in Chapter 2 is to reconfigure the critical questions as statements that can be classified as ordinary premises, assumptions or exceptions (Gordon and Walton, 2009). If we look at the list of critical questions for argument from correlation to cause (Section 3), it is possible to classify the first one as an ordinary premise. The next premise can be classified as an assumption. With this type of critical question, when the questioner merely asks the question, that is enough to shift the burden of proof to the other side to respond appropriately to the question, or else the argument fails. With the exception type of critical question, in order to shift the burden of proof to the other side, the questioner must present some evidence to back up the question if the proponent of the original argument demands such evidence.

Looking at Figure 7.6 we can see how the argumentation is structured in this case. The node contains the scheme for argument from correlation to causation. The premise at the top is an ordinary premise, while the second premise from the top is an additional assumption. Thus the two premises at the top are classified as assumptions. The remaining four premises are represented in Figure 7.6 as exceptions. This means that the four premises at the bottom are assumed not to hold, and only if evidence is given to

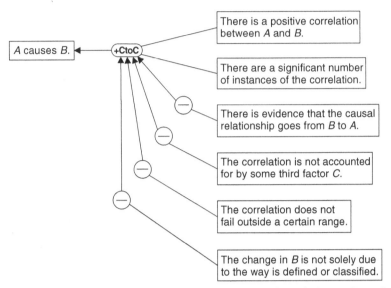

FIGURE 7.6 How Carneades Visualizes Argument from Correlation to Causation

back one of them up will the critical question defeat the argument from correlation to cause. In other words, the four premises at the bottom are displayed as undercutters, with contra arrows going to the argument node.

Evaluating the argumentation in a case like the pernicious anemia example is a matter of adjusting burdens of proof set and adjudicated at different stages of the dialogue. During the argumentation stage, as speech acts are put forward by both parties, and as arguments are brought forward and challenged by the other side, these arguments need to be evaluated in light of how the burden of proof shifts from one side to the other. Rules governing the allocation of burden of proof during the argumentation stage are set by the protocols for each type of move. For example, if a claim is made, and the other party challenges the claim, the first party has to support its claim with an argument or else he or she has to retract the claim. What is also necessary to take into account, however, is that there is a global burden of proof set at the opening stage, including a standard of proof that determines how strong the argumentation on each side needs to be in order to be successful in making its case in the dialogue.

The appropriate standard for the burden of proof for the statement to be proved or disproved that needs to be to be set at the opening stage of an inquiry is the beyond reasonable doubt standard. In contrast, in a discovery dialogue, no burden of proof is set at the opening stage. During the argumentation stage, as noted above, there is a different type of burden of proof that operates only during the argumentation stage that is called the burden of production. The two types of dialogue are not different in this respect during the argumentation stage, where arguments meeting any of the four standards can be put forward, considered, questioned, supported and accepted or rejected, based on any of the four standards. However, it is to be expected that arguments meeting only the first or second standard would be more common in a discovery dialogue, whereas arguments meeting only the third or fourth standard would be more common in an inquiry dialogue.

However, a qualification is necessary. There is a standard of proof that is applied to sort out which hypothesis should be accepted when the closing stage of a discovery dialogue is reached. In scientific argumentation, hypotheses are constructed during the argumentation stage, hypotheses that represent ways of solving the puzzle or problem posed at the opening stage. In a productive discovery dialogue, as the evidence comes in, it is shown that some hypotheses are better supported by the evidence than others, and even that some are refuted by the evidence. One may even emerge as the "best" hypothesis, the one supported by the most evidence and least open to attacks that might refute it. The standard used to select the best hypothesis is comparable to the preponderance of the evidence standard, except that it applies not to just to a single statement, but to a comparison between any pair of statements as a set. Of any pair, the one supported by more evidence, and rebutted by less evidence, is the one to accept. Thus there is a kind of burden of proof set at the opening stage of a discovery dialogue, as there is in an inquiry dialogue, except that it does not apply to a single statement set as the thesis to be proved or disproved at the opening stage. It applies in advance only to whatever statements may eventually emerge during the argumentation stage as hypotheses are invented.

10. Conclusions

The argumentation in the classic case of the investigation of pernicious anemia naturally breaks down into two different kinds of procedures for arriving at a conclusion by collecting, testing and evaluating evidence in a scientific investigation. During the procedure in the discovery dialogue, the problem posed was how to find a cure for a deadly disease. The investigators began by trying to solve the problem posed by the large numbers of people killed every year by the disease. They began by building on previous research that had found that anemic dogs could be treated by feeding them liver. When they tested this treatment on human patients, they found that it worked there, too. They also observed an increase in the red blood cell counts of these patients. They had found a practical way of treating this deficiency by feeding patients liver. Based on their observations of the correlation between feeding the patients liver and the subsequent return to health of the patients, argument from correlation to cause was the warrant for drawing an inference to a causal hypothesis between these two sets of events. These results were extremely useful in suggesting treatments for anemia. They led naturally to further experimental investigations and research.

The question at this point was transformed into the more theoretical one of determining precisely which factor was the cause of the increase in the blood cell count and subsequent return to health of patients affected by pernicious anemia. They had found a cure for the disease, through a process of discovery, and what they had found supported the hypothesis that pernicious anemia was caused by a deficiency of red blood cells. The need to better understand the link between cause and cure opened the need for an inquiry.

Although this finding was an excellent basis for treating patients, it did not yet meet a high enough standard of proof for drawing the conclusion as the finding of a scientific inquiry into the cause of pernicious anemia. It was only later experiments, showing that something in the liver reacted with some chemicals in the stomach, that led to further research finding that a deficiency of vitamin B12 was the cause. Only at this stage, when this chemical reaction was studied, and an explanation of it was given at an adequate theoretical level, could it be said that all the critical questions had been answered in such a way that it could be said to have been proved conclusively what the cause of the disease was and what the scientifically adequate explanation of it was. The structure of the procedure used to process the evidence that was gradually being accumulated had now shifted to that of an inquiry dialogue, where the need was that of collecting enough clinical and scientific evidence to prove that deficiency of vitamin B12 was the cause of pernicious anemia.

The methodology adopted in this chapter was mainly that of a top-down approach in which a formal and computational model of evidential argumentation was applied to the specific problem of understanding how a scientific investigation works as a process of inquiry along the lines suggested by the philosophies of science of Peirce and Popper. Traditional theorists in the philosophy of science had drawn a bright line between two stages, a discovery stage and later proof or justification stage. The recent climate of opinion in philosophy of science seems to be going away from the view that there is a sharp distinction between these two stages and toward the view that there is a continuous sequence of argumentation from the one stage to the other. The conclusion of this chapter seems to support the latter view more strongly, but also accommodates the former view to some extent, because it does postulate that there two different types of procedures for the collection and evaluation of scientific evidence. Strong support is given for the latter view, however, on the grounds that the Carneades model represents the idea that the first type of dialogue is embedded in the second one in cases where an initial discovery dialogue is the basis for moving the investigation forward to an inquiry stage where the conclusion can be proved to a higher standard. Although we have taken one step in the direction of providing a bottom-up analysis by applying the model to a single case, clearly the model will need more refinement when applied to more sophisticated cases of embedded scientific discoveries and inquiries.

Many more detailed case studies of scientific discovery and inquiry are needed to fully substantiate the model for scientific argumentation put forward in this chapter and to bring out the finer details of the embedding. In this model, the sequence is broken down into two main components, a discovery dialogue and an inquiry dialogue. Each of these components is represented by a different model of dialogue, and the sequence of argumentation in the case as a whole is modeled as a dialectical shift from one type of dialogue to another. The first phase is represented as a discovery dialogue, and the second as an inquiry dialogue.

While the same kinds of arguments might be used in both a discovery dialogue and an inquiry dialogue, what marks the difference between the two especially is that each has different burdens and standards of proof. The burden of proof to prove or disprove any specific statement is not set at the opening stage of a discovery dialogue. However, the standard appropriate for argumentation in a discovery dialogue that applies during the argumentation stage is that of preponderance of the evidence, meaning in the case of two hypotheses that the one with the greater weight of evidence supporting it is the one that should be accepted. However, in many examples of scientific discovery there may be more than two hypotheses. In this kind of case, the hypothesis with the greatest weight of evidence supporting it should be accepted. In either event, in the case of a tie, practical factors could come in to play, such as the comparative costs of conducting the experiments required to test the two hypotheses. The aim of the scientific discovery dialogue is to select the hypothesis most suitable for further testing and investigation. In an inquiry dialogue, in contrast, the aim is to terminate the investigation by proving the statement being investigated so conclusively that reopening the inquiry will turn out not to be necessary. Fulfilling this aim requires a higher standard of proof than preponderance of the evidence. An explanation that is to be acceptable as a proof of a causal hypothesis must meet a standard of proof such that it has to be better than competing explanations by a threshold appropriate for the field of investigation.

8

Fallacies, Heuristics and Sophistical Tactics

The purpose of this chapter is to advance fallacy theory beyond its current state of development by linking it to the notion of defeasible reasoning. Defeasible reasoning has turned out to be very important for computing, especially in view of the attention paid to modeling argumentation (Bench-Capon and Dunne, 2007) and the use of argumentation schemes (Verheij, 2003) and dialogues (Prakken, 2000; 2006) to study problems of nonmonotonic reasoning. The advent of argumentation frameworks (Dung, 1995) can be shown to provide an elegant way of subsuming much previous work on defeasible reasoning (Bondarenko, Dung, Kowalski and Toni, 1997). Although much has been written on individual fallacies, there is comparatively little on the general theory of fallacy, except for the pragma-dialectical theory (van Eemeren and Grootendorst, 1992) and the pragmatic theory (Walton, 1995). Defeasible reasoning is uncontestably important for helping us to better grasp the notion of fallacy and rethink it as a concept useful for modern logic, but so far the link between the two notions has not been studied.

Many of the most common forms of argument associated with major fallacies, such as argument from expert opinion, *ad hominem* argument, argument from analogy and argument from correlation to cause, have now been analyzed using the device of argumentation schemes (Walton, Reed and Macagno, 2008). Recent research in computing has embraced the use of argumentation schemes and linked them with key logical notions such as burden of proof that are also related to the study of fallacies (Gordon, Prakken and Walton, 2007). Argumentation schemes have been put forward as a helpful way of characterizing structures of human reasoning, such as argument from expert opinion, that have proved troublesome to view deductively. Attempting to deduce the reasonable examples, by viewing the major premise as a conditional not subject to exceptions (e.g., if X says Y, then Y is true) does not work at all well, as this type of argument is typically defeasible. As noted in the introductory section of this chapter, the traditional logic textbooks treated these forms of argumentation only under the heading of informal fallacies, but the trend now is to recognize that they are often useful arguments that we could not get by without relying on in many cases, even though they are dangerous because they are used fallaciously in some cases (Hansen and Pinto, 1995; Walton, 2006c). In this chapter, evidence is given to suggest that argumentation schemes and formal dialogue models are the tools needed for studying the properties of defeasible reasoning and informal fallacies.

Section 1 outlines the two leading theories of fallacy. Section 2 shows that twelve of the traditional major informal fallacies correspond to argumentation schemes, while seven of them do not. Section 3 gives a quick summary of the main features of defeasible reasoning that are especially important for the study of fallacies. Section 4 surveys default logics used to model defeasible reasoning in artificial intelligence. Section 5 explains how certain properties of dialogue systems are important for analyzing defeasible reasoning and informal fallacies. In Section 6, the profile of dialogue tool is applied to the fallacy of many questions to show how dialogue models can be applied to real instances of argumentation treated under the category of a traditional informal fallacy. Section 7 formulates dialogue conditions that connect defeasibility to fallacies and that reveal underlying features that can be used to analyze sequences of reasoning associated with fallacies. Section 8 takes as a case in point the fallacy most closely related to defeasible reasoning, *secundum quid* (the fallacy of neglect of qualifications to a general rule). Section 9 features another case in point, the defeasible lack of knowledge type of argument associated with the traditional fallacy of *argumentum ad ignorantiam*. Section 10 redefines the notion of fallacy and draws conclusions from what has been shown in the chapter.

1. Theories of Fallacy

The most fully developed theories of fallacy so far are the pragmatic theory and the pragma-dialectical theory of the Amsterdam School (Tindale, 1997). The two theories have much in common, but define the notion of fallacy in different ways. According to the pragmatic theory (Walton, 1995, 237–238), there are six basic characteristics of fallacy. These characteristics state conditions on what reasonably should be taken to be a fallacy according to the criteria discussed in Walton (1995).

1. A fallacy is a failure, lapse or error, subject to criticism, correction or rebuttal.
2. A fallacy is a failure that occurs in what is supposed to be an argument.
3. A fallacy is associated with a deception or illusion.
4. A fallacy is a violation of one or more of the maxims of reasonable dialogue or a departure from acceptable procedures in that type of dialogue.

5. A fallacy is an instance of an underlying, systematic kind of wrongly applied technique of reasonable argumentation.
6. A fallacy is a serious violation, as opposed to an incidental blunder, error or weakness of execution.

These six basic characteristics set in place requirements that any theory should meet, or at least should be prepared to cope with, in order to be a satisfactory theory of fallacy.

The fallacy of many questions arguably fails to meet condition 2, because the speech act of asking a question is different from the speech act of putting forward an argument. For example, a complex and loaded question such as, 'Are you confused when you are not on your medications?' has traditionally been taken to fall under the category of the fallacy of many questions even though, at least apparently, it does not have the form of an argument. What needs to be said here are two things. First, this question implicitly contains an argument, and indeed that is partly what is fallacious about it. The argument is concealed. Second, it uses an argumentation strategy of attempting to take three turns at once in a dialogue, instead of following the proper rule of taking only one turn at a time.[1]

The leading competitor to the pragmatic theory of fallacy (Walton, 1995) has so far been the evolving doctrines of the Amsterdam School. According to the earliest version of their theory, a fallacy is a violation of a rule of a critical discussion (van Eemeren and Grootendorst, 1992). This theory was a good advance at the time because it went beyond the older treatments of the logicians who so often tended to analyze a fallacy by seeing it only as a failed inference of some sort. It was a breath of fresh air. However, it was argued in Walton (1995) that this theory is still too narrow for several reasons, and these reasons will be illustrated and amplified below using a few examples. First, as shown in Section 2 below, many if not most of the traditional fallacies are associated with argumentation schemes that represent forms of inference of various kinds. Both the inferential aspect and the dialectical aspect of the notion of a fallacy need to be taken into account. But the Amsterdam theory is also too narrow for other reasons cited in Walton (1995, 298). Violating a rule of a critical discussion should not be itself equated with the committing of a fallacy, for some such violations are merely blunders and not fallacies. Another problem with their theory is that many different fallacies can be analyzed as different ways of violating the same rule. Still another problem, as will be illustrated below, is that in order to properly analyze many of the fallacies one has to examine the argumentation strategy on which the fallacy was based. The theory has been more recently strengthened by the work of van Eemeren and Houtlosser (2006) on strategic maneuvering, and even further by van Eemeren (2010).

---

[1] This issue will be discussed further in Section 5.

The following short definition can be given, based on the pragmatic (Walton, 1995) theory of fallacy: a fallacy is an argument, a pattern of argumentation or something that purports to be an argument that falls short of some standard of correctness as used in a conversational context but that, for various reasons, has a semblance of correctness about it in context and poses a serious obstacle to the realization of the goal of the dialogue. A pattern of argumentation is an ordered sequence of moves by two parties in a dialogue. This theory is inherently dialectical in that not only is the structure of the inference from the set of premises to the conclusion taken into account in evaluating argumentation in a particular case, but whether or not an argument is fallacious in that case also needs to depend on the context of how it was used in a sequence of moves in a dialogue.

Lewinski (2011) has observed that because fallacies are often arguments used in a deceptive way that makes them difficult to distinguish from reasonable argumentation modes, contextual variation has to be taken into account. He cites the example (480–481) of the informal fallacy of argument from popular opinion, or appeal to the people, as it is sometimes called. His example is the argument "we should do what the majority of the people want". This move could be a reasonable argument in a democratic debate, even though many examples of appeal to the people are often cited logic textbooks as instances of the fallacy of *argumentum ad populum*. Appeal to the people is generally a weak kind of argument, and one that can be misleading and go badly wrong in situations where better evidence is required to prove a claim. As with many arguments associated with fallacies, an inconclusive and defeasible argument can justify acting on a tentative presumption but can also go badly wrong when too much weight is placed on it or when it is used deceptively as a device to conceal a failure to collect enough evidence to prove a point properly.

Many of the fallacies are fallacious moves in a dialogue not because of the inherent unreasonableness of the argument but because of the way it is used in a sequence of moves to try to prevent the respondent from questioning it or even continuing the dialogue at all. The classic case of this type of fallacy is the fallacy of poisoning the well. For example, in the Cardinal Newman case, the attack alleged that as a Catholic, Newman could have no regard for the truth of the matter in any political discussion (Walton, 2006c). In this case, as reported by Copi and Cohen (1994, 124) the novelist Charles Kingsley attacked the Catholic intellectual John Henry Cardinal Newman by arguing that Newman's first loyalty could not be to truth, implying that Newman's claims were not to be trusted. The argument offered was that Newman's first loyalty could not be to truth, because, as a Roman Catholic priest, his prior loyalty always had to be to the Catholic Church. Newman's response was that this *ad hominem* attack made it impossible for him, or even for any Catholic for that matter, to have any credibility in public argumentation. According to Newman, Kingsley had "poisoned the well" against whatever Newman might say. Or to cite another excellent example, an appeal to expert opinion might cite a scientific authority and dismiss any reasonable attempt to ask critical questions about the argument from expert opinion by declaring that any evidence appearing to go against it must be dismissed as anecdotal. The fallacy in both instances is found not in the argumentation scheme, as applied to a single argument, but in a pattern that can be found only by examining a connected sequence of moves by both parties.

On this dynamic approach, a distinction has to be drawn between two kinds of fallacies. In some cases, a fallacy is merely a blunder or an error, while in other cases, it is a sophistical tactic used to try to get the best of a speech partner in dialogue unfairly, typically by using verbal deception or trickery. The evidence of the use of such a tactic is found in the pattern of moves made by both sides in the dialogue. It is important for fallacy theory to avoid being impaled on the horns of a dilemma between these two traditional types of problematic argumentation moves. To confront the dilemma, the pragmatic theory (Walton, 1995) distinguished between two kinds of fallacies. The paralogism is the type of fallacy in which an error of reasoning is committed typically by making a blunder by failing to meet some necessary requirement of an argumentation scheme. The sophism is a sophistical tactic used to try to unfairly try to get the best of a speech partner.

Once we realize that fallacies are associated with defeasible argumentation schemes, there can be recognition that many of the types of arguments generally presumed to be fallacious, such as argument from expert opinion in the traditional accounts on fallacies in logic textbooks, are basically reasonable arguments that have been used wrongly. Cognitive science studying heuristics and cognitive biases in human decision-making has recognized that argumentation typically depends on quick rules of thinking that enable us to rapidly solve a problem under conditions of uncertainty. According to the dual-processor theory there are two kinds of thinking. One is calculative, conscious and slow. The other is unconscious and fast, and even though it may tend to jump to a conclusion too quickly in some instances, it is extremely useful when making a fast decision under constraints of time pressure and lack of knowledge. Such heuristics are used, for example, in medicine where a physician needs to decide under time pressure whether a patient admitted to the emergency room should be classified as low risk or high risk (Gigerenzer, Todd and the ABC Research Group, 1999, 4–5). A quick calculation made on the basis of the immediately available data of blood pressure, age and cardiac rhythms can be used as a heuristic to decide where to send the patient next.

Consider once again the classic case of argument from expert opinion, typically treated in the traditional logic textbooks under the heading of appeal to authority, implying that arguments based on expert opinion are subjective and are no substitute for collecting objective data. Now, however, it is widely recognized that we often need to depend on expert opinions, especially in public deliberations and in making judgments based on evidence in law. So with good reason, it has come to be accepted gradually that this older presumption is unsustainable. It is better to see argument from expert opinion as based on a fallible heuristic that can be expressed in the generalization that if an expert says a particular proposition is true, that is good reason, subject to exceptions should we find reasons to think otherwise, that the proposition can at least tentatively be expected as true as the basis for moving forward in a dialogue. Such a general principle can be called a parascheme (Walton, 2010b), a device representing the structure of the heuristic as a rapid but defeasible form that can be used to derive a conclusion tentatively subject to retraction as new evidence comes to be considered.

Once again, using the example of argument from expert opinion, the parascheme takes the two ordinary premises (that E is an expert and that E asserts that A) and uses them to directly infer the conclusion that proposition A is tentatively acceptable. As we know, however, from studying the structure of the argument from expert opinion in the preceding chapters, this inference from only the ordinary premises overlooks the assumptions and exceptions (modeled in Carneades) that are additional premises that need to be taken into account because of their role as critical questions. There are other factors that need to be considered. We have to ask whether the expert is knowledgeable in the right field, whether his or her assertion is based on evidence and so forth. But because a heuristic needs to arrive at a fast, tentative conclusion on a defeasible basis, such critical questions can be momentarily set aside at an early stage in the dialogue in forming a hypothesis. For as the dialogue proceeds, this temporary hypothesis can be evaluated more fully at a later stage as more relevant evidence pro and contra comes in.

From the point of view of logical argumentation, problems come in when the need to ask such critical questions is ignored or, even worse, where the proponent of the argument from expert opinion insists on requiring the respondent not to even think of asking critical questions. For example, if the respondent attempts to ask critical questions, the proponent might try to shield such attempts by saying something like, "You are not an expert, so you are not entitled to say anything about this at all." Once we start to examine examples of arguments classified under the heading of 'fallacy of appeal to authority' in logic textbooks, we see that there are two kinds of fallacies involved. One is the error of simply leaping ahead too quickly to a conclusion. The other is the fallacy of trying to put up a shield to prevent the respondent from asking any critical questions at all. On this basis it is possible to classify fallacies into the two categories specified above. One type is the paralogistic error of jumping to a conclusion without properly considering the critical questions that should be asked at that point in the dialogue. The other may be a little more difficult to diagnose, because it involves looking at strategic maneuvering in a more lengthy sequence of moves in a dialogue. In the case of the sophistical tactics of fallacy there is a pattern on the part of the proponent of the argument to prevent the respondent from properly asking critical questions using various tactics. The study of such tactics requires looking at a pattern of moves by the one party and responses by the other party consisting of the sequence of speech acts in the dialogue, based on the normative model of dialogue appropriate for the case. As van Eemeren (2010, 198) puts it, derailments of strategic maneuvering can occur in cases where commitments to having a reasonable exchange are overruled by the need to put forward an argument that will be effective to cleverly persuade an audience.

Some would say that a fallacy is an intentional deception, distinguishing between the two kinds of fallacy by saying that the sophistical tactics fallacy is an intentional deception, while the paralogistic one is a mere mistake. This way of drawing the line misses the point, however. Many fallacies are committed because the proponent has such strong interests at stake in putting forward a particular argument, or is so fanatically committed to the position advocated by the argument, that he or she is blind to weaknesses in it that would be apparent to others not so committed. In this kind of case the deception may not be intentional, because the proponent does not see the argument as faulty. The proponent is so committed to this kind of argumentation that he or she pushes ahead with it blind to errors that others might find in it. But intention is an internal mental concept that can be inferred only abductively based on the external evidence of what the agent knew or considered. But whether it is intentional or unintentional does not really matter from a point of view of analyzing the argument and deciding whether it should be considered fallacious. What is important from a viewpoint of logical argumentation is the logical weakness in the argument, or some fault in the pattern of argumentation, not some psychological fault in the arguer.

The sophistical tactics type of fallacy tends to be a more serious kind of problem than the error of reasoning one. It is based on the idea that an organized rule-governed dialogue in which arguments are exchanged, like a critical discussion, is partly adversarial but also partly cooperative. Collaborative procedural rules are very important in such a dialogue, but there is also an adversarial element. A participant in a critical discussion is an advocate of his or her own viewpoint and a critic of the opposing viewpoint. Thus a critical discussion is like a free market economy in which each side tries to win by having the strongest argument that will triumph over those of its opponents. The problem is that bad things can happen, including the committing of fallacies by shifting to a quarrel that is purely adversarial. Any conversation in which reasonable argumentation is to be used for some constructive purpose must strike the right balance between this adversarial aspect and the need for all parties to follow the Gricean maxims of polite conversation that are required to make the contributions of each participant useful to move the dialogue toward its goal. It is here that strategic maneuvering (van Eemeren, 2010) provides a helpful tool to analyze cases where a sophistical tactic has been committed.

Another problem is that to analyze fallacies properly, we have to explain how each of them is used as an effective deceptive tactic that does work to fool people. The theory of strategic maneuvering (van Eemeren, 2010) is the best tool for this task because it can take the strategic dimension of fallacies into account. Strategic maneuvering refers to the efforts of arguers in a discussion to reconcile their twin aims of rhetorical effectiveness and maintaining dialectical standards of reasonableness. These twin aims arise from the fact that the critical discussion has both an adversarial and a collaborative component. Participants have to collaborate by following the rules for the critical discussion, for example, by taking turns putting forward arguments. At the same time, however, each side is trying to convince the other side to accept its point of view by using the strongest arguments possible to support its thesis and to refute the thesis of the opposed side. This dual aspect of the critical discussion means there is always a tension between rhetorical effectiveness and using arguments that fit dialectical standards of reasonableness. The critical discussion is like a free market economy in which people have to follow laws and social rules, but within the framework of these rules they are free to try to maximize their own profits. In such a competitive situation, an arguer is free to use strategic maneuvering to try and put forward a winning argument. It is up to the other side to ask critical questions and generally to probe into an argument to try to find the weak points in it, and if possible to refute it by counterarguments. A burden of proof is placed on one side or the other to fulfill its dialectical obligations.

Even over and above strategic maneuvering, however, participants in a critical discussion tend to have interests at stake. In a philosophy discussion about some theoretical subject in a classroom, participants who are students might want to show that they are better than the other students by using strong but fallacious arguments to try to refute the arguments of their opponents. Moreover, they might want to impress their professor in the hope of showing how clever they are, thinking that this might help them get a better grade in the course. What these observations show is that interests are involved outside the framework of the critical discussion itself. These interests relate to the goals of the participants outside the critical discussion and how they might use their arguments in the critical discussion for some external purpose. What this may suggest is that the explanation for committing fallacies may be found in the observation that the participants involved in a critical discussion are also involved in other goal-directed types of activities at the same time.

2. Argumentation Schemes and Fallacies

Many of the most common forms of argument associated with major falla-cies, such as argument from expert opinion, *ad hominem* argument, argu-ment from analogy and argument from correlation to cause, have now been analyzed using the device of argumentation schemes (Walton, Reed and Macagno, 2008). We need to recall that although the traditional logic textbooks mainly treated these forms of argumentation under the heading of informal fallacies, in many instances, they are reasonable but defeasible arguments. The formal and inductive fallacies, such as affirming the con-sequent and arguing from too small a sample to a generalization, can be analyzed with deductive and inductive forms of reasoning familiar to for-mal logicians. However, of the major informal fallacies, the following twelve need to be analyzed with defeasible argumentation schemes of the sort that can be found in Walton, Reed and Macagno (2008, chapter 9).

1. *Ad Misericordiam* (Scheme for Argument from Distress, 334)
2. *Ad Populum* (Scheme for Argument from Popular Opinion and its subtypes, 311)
3. *Ad Hominem* (*Ad Hominem* Schemes; direct, circumstantial, bias, 336–338)
4. *Ad Baculum* (Scheme for Argument from Threat, 333; Fear Appeal, 333)
5. Straw Man (Scheme for Argument from Commitment, 335)
6. Slippery Slope (Slippery Slope Schemes; four types, 339–41)
7. *Ad Consequentiam* (Scheme for Argument from Consequences, 332)
8. *Ad Ignorantiam* (Scheme for Argument from Ignorance, 327)
9. *Ad Verecundiam* (Scheme for Argument from Expert Opinion, 310)
10. *Post Hoc* (Scheme for Argument from Correlation to Cause, 328)
11. Composition and Division (Argument from Composition, 316; Division, 317)
12. False Analogy (Scheme for Argument from Analogy, 315)

Of course, these are not the only fallacies that can be analyzed with the help of argumentation schemes. We include here for discussion only the major falla-cies that are most commonly treated in the most widely used logic textbooks.

An example is the scheme for argument from expert opinion with its matching set of critical questions (Walton, Reed and Macagno, 2008, 310). This scheme represents a defeasible form of argument that is not well mod-eled as deductive or inductive. Instead, it is viewed as an argument that can hold tentatively under conditions of lack of knowledge of the full facts of a case but that can be defeated or cast into doubt by the asking of appropri-ate critical questions. On this view, argument from expert opinion can be a reasonable argument in some instances of its use, provided it is realized that it is a defeasible heuristic that can be given up if more facts come to be known about the case so that we no longer need to rely on expert opinion. In other instances of its use, however, it can be fallacious. For example, if the proponent of the argument treats it as infallible and refuses to concede that it is even open to critical questioning, that would be a fallacious misuse of the argument.

Arguments such as appeal to expert opinion are tricky because they are often very useful, and in many instances they are the best resources we have for navigating through a world where time and resources are limited, but where a decision to accept a hypothesis or course of action needs to be taken. Argument from expert opinion needs to be seen as open to critical questioning in a dialogue. You can treat it as a deductively valid argument or as an inductively strong argument in some instances, but in the vast majority of cases, this way of treating it would lead to serious problems (even fallacies). Generally, it is not a good policy when examining expert witness testimony to assume that the expert is omniscient. Indeed, such a policy would be highly counterproductive when evaluating this kind of evidence in a common law trial. However, there is a natural tendency in everyday reasoning to respect expert opinions and even to defer to them or to hesitate to question them. Questioning the opinion of an expert can seem impolite, unless done in a circumspect way. However, as a matter of fact, experts are often wrong, or what they say can be highly misleading, leading to the drawing of a conclusion that is wrong or not supported by the evidence. As a practical matter, one often needs to be prepared to critically question the opinion of an expert by asking the right questions. Such an argument needs be to seen as defeasible, or subject to default.

Tindale (2007) agrees that many fallacies are misuses of argumentation schemes that are legitimate but defeasible argument strategies, citing *ad hominem* and *ad verecundiam* as leading cases in point. However, he cites the straw man as an example of a fallacy that does not fit into this category. He writes (12) that there is no way we can make it fit "unless we conjure up something trivial such as Real Man". But as Krabbe (2007, 129) noted, "Real Man is not so trivial, considering how hard it is to correctly explicitize the implicit elements in one's opponent's position". Straw man, Krabbe adds, can be seen as a derailment of such explicitization strategies.

As well, the straw man fallacy can be seen as a derailment of argument from commitment. The scheme for argument from commitment (version 1) from Walton, Reed and Macagno (2008, 335) is given below.

Commitment Evidence Premise: In this case it was shown that a is committed to proposition A, according to the evidence of what he said or did.

Linkage of Commitments Premise: Generally, when an arguer is committed to A, it can be inferred that he is also committed to B.

Conclusion: In this case, a is committed to B.

The following are the two critical questions matching argument from commitment (Walton, Reed and Macagno, 2008, 335).

> **CQ$_1$:** What evidence in the case supports the claim that *a* is committed to *A*, and does it include contrary evidence, indicating that *a* might not be committed to *A*?
>
> **CQ$_2$:** Is there room for questioning whether there is an exception, in this case, to the general rule that commitment to *A* implies commitment to *B*?

The fallacy of straw man occurs where the first party in a dialogue has distorted what is taken to the second party's implied commitment to *B*, making the attributed commitment appear more implausible or more extreme, and thereby more open to attack. In any given case however, this fallacy cannot be (completely) pinned down by using the argumentation scheme for argument from commitment. Matters of the context of dialogue need to be taken into account. We need to examine carefully what each party said in relation to what the other party claimed he or she said, and how he or she used that in the argument.

Note however that version 2 of the scheme for argument from commitment given in Walton, Reed and Macagno (2008, 335) does take dialogue factors into account.

> **Major Premise:** If arguer *a* has committed him- or herself to proposition *A*, at some point in a dialogue, then it may be inferred that he or she is also committed to proposition *B*, should the question of whether *B* is true become an issue later in the dialogue.
>
> **Minor Premise:** Arguer *a* has committed him- or herself to proposition *A* at some point in a dialogue.
>
> **Conclusion:** At some later point in the dialogue, where the issue of *B* arises, arguer *a* may be said to be committed to proposition *B*.

This version of the scheme, however, is not sustainable, for the following reasons. If *A* entails *B* and a participant is committed to *A*, then it may be inferred that he or she is also committed to *B*, whether or not *B* is an issue. In other words, whether or not *B* has become an issue at some later point in the dialogue should be irrelevant for the purpose of an argumentation scheme. Evidence for this is found in the need for consistency with the formulation of other argumentation schemes. No other argumentation scheme restricts the conclusions that may be drawn specifically to conclusions that are relevant for issues that have been raised in the dialogue. Such a relevancy condition is not suitable to be placed within an argumentation scheme. Its proper setting is that of dialogue rules.[2] The lessons of

---

[2] I would like to thank an anonymous referee for bringing this important point to my attention.

these points are significant, for it has been shown that while the scheme for argument from commitment is necessary for the determination of instances of the straw man fallacy, it is not sufficient. Dialogue factors also need to be taken into account.

Next we need to see that there are at least seven major fallacies remaining that do not fit any of the argumentation schemes: (1) equivocation, (2) amphiboly, (3) accent, (4) *petitio principii* or begging the question, (5) *ignoratio elenchi* (irrelevance: species are red herring and wrong conclusion), (6) *secundum quid* (neglecting qualifications) and (7) many questions. The three linguistic fallacies – equivocation, amphiboly and accent – are not based on specific argument types. They have to do with ambiguous communications (speech acts) in dialogues. Begging the question occurs in circular chains of reasoning where the links in the chain can consist of many different kinds of arguments. Irrelevance is failure to prove a specified conclusion that is supposed to be proved. Many kinds of arguments can commit this error. The error is not specific to the type of argument used. Irrelevance and begging the question are fallacies that have to do with sequences of extended argumentation where the links of inference making up the reasoning can be of different kinds, including any types represented by the various argumentation schemes. The fallacy of *secundum quid* is not specific to a type of argument, unless it is perhaps arguing from a generalization to an instance or arguing from an instance to a generalization. But such inferences can occur in any kind of defeasible argumentation. The fallacy of neglecting qualifications occurs in any such instances where proper qualifications are ignored or suppressed, as will be shown in Section 5. However, it should be noted that there are a number of argumentation schemes in which this fallacy plays an important role. As noted above, the fallacy of many questions is identified not with a specific type of argument, but rather with a strategy of questioning in a dialogue format. Hence none of these seven fallacies is a misuse of any particular argumentation scheme.

3. Defeasible Reasoning

The etymology of the term 'defeasible' comes from medieval English contract law, referring to a contract that has a clause in it that could defeat the contract in a case where the circumstances fit the clause. However, the origin of the term in modern philosophy and law is a chapter called 'The Ascription of Responsibility and Rights' by H.L.A. Hart (1949). Hart's work was attacked in subsequent years by philosophers who criticized it heavily (Loui, 1995, 21), even though a few, such as Toulmin, accepted and used it. But his view turned out to be prescient, in light of the importance defeasible reasoning turned out to have in computing. As noted in the introductory section above, many formal systems of defeasible reasoning were produced in the fields of artificial intelligence and logic. There are several different approaches. Some place defeasible reasoning in a context of new information coming in that annuls a previous conclusion drawn by inference. Others see defeasibility as operating in a framework of belief revision, as an agent updates his or her beliefs. Hart saw it as circumstances fitting an exception to a general rule.

The originating idea behind Hart's way of defining the term can be appreciated from Hart's work of 1949, quoted from Loui (1995, 22).

Claims can usually be challenged or opposed in two ways. First, by a denial of the facts upon which they are based and secondly by something quite different, namely a plea that although all the circumstances on which a claim could succeed are present, yet in the particular case, the claim ... should not succeed because other circumstances are present which brings the case under some recognized head of exception, the effect of which is either to defeat the claim ... altogether, or to "reduce" it....

Judging from this quotation, it would appear that Hart had the idea of a claim being at first acceptable because it is supported by reasoning but is later defeated because circumstances are present that bring the case under an exception. Thus we recognize the idea of a defeasible argument, of a kind so common in law.

However, that is not the only way Hart saw defeasibility. He also discussed defeasible concepts. His most famous example is from *The Concept of Law* (1961). Consider the rule that no vehicles are allowed in the park. This rule could be defeated by special circumstances, for example, during a parade, but it could also be defeated because of the open texture of the concept of a vehicle. For example, a car would definitely be classified as a vehicle, and be excluded from the park, but what about a bicycle? Is it a vehicle? Both sides could be argued, unless the law makes a specific ruling on bicycles. The literature on computing has concentrated on defeasibility of arguments rather than on defeasibility of concepts,[3] and these two notions seem to be quite different. However, the precise distinction between these two kinds of defeasibility needs to be clarified. The notion of a defeasible concept presupposes that the concept already has some definition, but the existing definition turns out to be inadequate in some new cases where it is unclear whether or not an entity in the case fits the definition. The most obvious way to handle this problem is to redefine the concept so that it is made clear whether or not the entity in the case fits the definition. But there is another way to deal with the problem. This second way is to bring forward

---

[3] This claim needs to be qualified. The HYPO line of work, initiated by Rissland and Ashley (1987), modeled defeasible reasoning with cases about the meaning of the open-textured term 'trade secret'. It should also be noted that logic programs can often be regarded as offering a set of definitions to work with, and these definitions are often defeasible. This is especially true in legal expert systems based on formalization of legislation, a well-known example being Sergot, Sadri, Kowalski et al. (1986).

a new set of inference rules that supplement the existing definition of the concept. Two examples of the second way of dealing with the problem of open-textured contexts in law are useful to consider.

The first is the case of the drug-sniffing dog (Weinreb, 2005). If a trained dog sniffs luggage left in a public place and signals to the police that it contains drugs, should this event be classified as a search according to the Fourth Amendment? If it can be classified as a search, information obtained as a result of the dog sniffing the luggage is not admissible as evidence. The problem is that although the concept of a search is partially defined in law, it may be open to contention whether this case fits the existing rules that provide the partial definition.[4] Weinreb (2005, 24) cited two rules established by prior court decisions that can helpfully be applied to the argumentation in the problem. One is the rule that if a police officer opens luggage and then observes something inside the luggage, the information collected is classified as a search. Another is the rule that if a police officer obtains information about a person or thing in a public place without intrusion on the person or taking possession of or interfering with the use of the thing, it is not a search for purposes of the Fourth Amendment. In the case of the drug-sniffing dog, the police officer did not open the luggage, so it can be argued on the basis of the first rule that what he or she did was not a search. Since there was no intrusion or interference, it can also be argued on the basis of the second rule that what he or she did should not be classified as a search.

In the case of *Popov v. Hayashi*, Barry Bonds made his record-breaking seventy-third home run in 2001 by hitting a ball into the stands where it was stopped by a fan named Popov in the upper webbing of his baseball mitt.[5] Before he was given the chance to complete the catch, he was thrown to the ground by a mob of fans trying to grab it, and when the melee was sorted out, another fan named Hayashi had secured possession of the ball. The case went to trial in the Supreme Court to decide which of these two fans can be said to have secured possession of the ball. Part of the problem in the case is the issue of whether Popov may properly be said to have caught the ball. Gray (2002), in his brief on the case, formulated some rules that are helpful in defining the notion of a catch. The first is the rule that a catch does not occur simply because the ball hits the fan on the hands or enters the pocket or webbing of the fan's baseball glove. The second is the rule that a catch does occur when the fan has the ball in his hand or glove, the ball remains there after its momentum has ceased, and even remains

---

[4] Here we need to be careful to note that many legal theorists do not consider precedent cases as defining legal rules. Reasoning with precedent cases is a theory construction process in which rules are hypothesized and then critically evaluated, and the *ratio decidendi* of a case is not legally binding.

[5] It may be interesting to note that the decision in the case of *Popov v. Hayashi* has been modeled by Wyner, Bench-Capon and Atkinson (2007) using a set of argumentation schemes.

there after the fan makes incidental contact with a railing, wall, the ground or other fans who are attempting to catch the baseball or get out of the way. Both rules provide foundations for arguments that Popov did not, properly speaking, catch the ball.

Cases like these can be used to show that there is a very close connection between the defeasibility of concepts as studied by Hart and the defeasibility of rule-based arguments of the kind studied in artificial intelligence. Indeed, it may even be suggested by a consideration of such cases that defeasible concepts can be reduced to defeasible arguments. Work on case-based reasoning in the field of artificial intelligence and law, for example, Rissland and Ashley (1987) and Costantini and Lanzarone (1995), has even attempted to model reasoning with open-textured concepts to reasoning with defeasible rules of the kind studied in mainstream artificial intelligence.

4. Default Logics

Default logic (Reiter, 1980) and circumscription (McCarthy, 1986) were developed around the same time to deal with nonmonotonic inference, and both formalisms have been extensively researched in computer science since that time. Both are designed for reasoning in the absence of complete information, where a tentative conclusion is drawn based on plausible assumptions needed to fill in missing details necessary to carry out an action or solve a problem. New incoming information may require the retraction of the conclusion so arrived at if it turns out that the assumption fails to hold once this new information comes in. One application of these formalisms is to cases of communication conventions. The following example was presented by McCarthy (1986, 3–4). Suppose I hire you to build me a birdcage and you fail to put a top on the cage. It would be ruled by a judge that I do not have to pay for the cage even though I had never explicitly said to you that my bird can fly. On the other hand, if I were to complain that you wasted money by putting a top on the cage that I intended for a penguin, the judge would rule that if the bird was of a kind that could not fly, I should have told you this before you commenced work on the cage.

In default logics of the kind used in artificial intelligence, first-order logic is extended with domain-specific rules called defaults (Reiter, 1980). A default rule has the following form, where P is a set of statements that act as given premises and D is another statement that could be called a default blocker. The form of such a default rule is: P, D; therefore C. A rule of this form tells you that if you know A, and you have no evidence of not D, then you may infer C. Another way to formulate a default rule is as a knowledge-based conditional of the following form: if you know that P is true, and you have no evidence that D applies, then you may infer C. In the case of the Tweety argument, A is 'Tweety is a bird', D is 'Tweety is not an exceptional bird' and C is 'Tweety flies'. As long as the default blocker applies, then the default rule works and the defeasible argument can be treated like any other deductively valid argument. Essentially, then, on this theory, a defeasible argument is analyzed as a default inference in which the warrant is a default rule. Immediately, the reader will recognize that both forms of defeasible reasoning bear a strong resemblance to the argument from ignorance in the list of fallacies in Section 2. This resemblance will be studied in greater detail in Section 8 on the argument from ignorance.

The problem with formal default logics is how you know in a given case whether the default blocker applies. We may not know, for example, that Tweety is an exceptional bird, but then later we may find out that he is. It is in the nature of many defeasible arguments that we do not know what lies in the future as new knowledge comes in. Indeed, according to the theory expressed by default logics, all defeasible arguments are arguments from ignorance. Thus in evaluating any given instance of such an argument, it depends on how far along an investigation has gone. If there is no evidence that D applies, an arguer can put forward a default argument and the respondent of the argument has to accept the conclusion, at least provisionally. But matters of burden of proof complicate such cases. We may think that Tweety is not an exceptional bird, for example, but if we are very worried that he might be, we might draw a different conclusion. Suppose Tweety has to carry an important message to military forces that depend on the information in the message. In such a case we might have doubts about how much weight we can put on the assumption that Tweety is not an exceptional bird, and look to also using other methods of sending the message. Such matters of burden of proof are very important for evaluating defeasible argumentation of the kind associated with fallacies. Prakken and Sartor (1996, 194) have modeled defeasible legal argumentation by using the notion of reversal of burden of proof. Defeasible arguments often have to do with presumptions that involve a reversal of the normal burden of proof.

Thus there are various distinctive aspects of defeasible arguments that suggest they are more complex than they seem. The default rule does indicate how they work but is limited in certain respects in explaining how they should be evaluated. The same default argument may be evaluated quite differently at a different stage in the procedure whereby new information is collected and arguments evaluated. At different stages of the process, defeasible arguments may vary in how they should be evaluated. The burden of proof may shift back and forth, depending on how far along an argument has proceeded.

In law, there are three kinds of burden of proof (Williams, 1977; Prakken and Sartor, 2009). The burden of persuasion is set before the beginning the trial, and never (or only rarely) shifts throughout the whole trial.[6] How persuasive such a winning argument needs to be depends on the standard of proof for that type of trial. In a criminal trial, the prosecution has to prove all the elements of the offense beyond a reasonable doubt, whereas in a civil trial the winning side must merely have a stronger argument than the losing side. In contrast, the evidential burden (also often called the burden of producing evidence or the burden of production) can shift from one side to the other (Fleming, 1961: Williams, 1977). The evidential burden refers to "the burden of producing evidence on an issue on pain of having the trial judge determine that issue in favor of the opponent" (Williams, 2003, 166). The burden of production refers to the quantity of evidence that the judge is satisfied with to be considered by the jury as a reasonable basis for making the verdict in favor of one side (Wigmore, 1940, 279). According to Williams (2003, 166) and Prakken and Sartor (2009, 228), there is also a third meaning of burden of proof. In this sense of the term, 'burden of proof' means that if the party "does not produce evidence or further evidence he or she runs the risk of ultimately losing on that issue". This third type of burden of proof involves a tactical evaluation of who is winning or losing at a particular point during the sequence of argumentation in the trial, and so Williams calls it the tactical burden, as opposed to the evidential burden. Gordon and Walton (2009) clarified the distinction between the burden of persuasion and the tactical and evidential burdens by showing that a trial has three stages: an opening stage, an argumentation stage and a closing stage. The argumentation stage can then be broken down into a sequence of smaller stages, where each smaller stage consists of all the arguments that have been put forward by both parties so far in the proceeding. The parties take turns in a dialogue putting forward arguments. The burden of persuasion is set at the opening stage and is used at the closing stage to determine which side won the trial. The two other burdens apply only during the argumentation stage. In some instances there can be meta-dialogues to argue about whether burden of proof should be changed or which side should have burden of proof (Walton, 2007b).

According to Gordon and Walton (2009) and Prakken and Sartor (2009), there are also two other types of burden of proof that can be distinguished in law. A person who feels he or she has a right to some legal remedy has the burden of claiming, that is, the burden of initiating the proceeding by filing a complaint alleging facts entitling him or her to some remedy. The burden of questioning requires that during pleading, an allegation of fact by either party is to be implicitly conceded unless it is denied.

---

[6] It seems to a bone of contention whether the burden of persuasion never shifts or whether it can sometimes shift but rarely does (Fleming, 1961, 62).

The notion of burden of proof is much better modeled, and makes more sense, in a dialogical framework rather than in the traditional more restricted logical framework in which an argument is merely seen as a set of reasons (premises) supporting a claim (conclusion). There were attempts to adopt a dialogical view of burden of proof in artificial intelligence in the early 1990s, for example, Bench-Capon, Lowes and McEnery (1991) and Gordon's influential pleadings game (1995). Now the dialogical approach has become fairly common in artificial intelligence and law. One serious motivation that brought argumentation theory into use in artificial intelligence was the need to conduct reasoning under conditions where the circumstances are changing and there is a lack of complete knowledge (Bench-Capon and Dunne, 2007, 621). The linkage between argumentation and defeasibility was seen in the adoption of the view in which an argument was taken to be a device for presenting a justification for a claim made, including the notion that an argument can be defeated as new information comes in. It seemed at this stage, and no doubt still does to many, that an argument should be viewed as a one-sided process in which a single party merely presents the reason that might be defeated by later evidence. However, in the more recent work on argumentation in artificial intelligence, there has been increasing recognition of argumentation as a dialogical process.

One of the driving forces behind the dialogical view of argumentation has been work in multiagent systems where dialogue was seen as a natural model of interaction between agents (McBurney and Parsons, 2002). Another is the need to study informal fallacies. The insight here (Bench-Capon and Dunne, 2007, 623–624) is that particular fallacies can be analyzed only by studying the given argument in relation to potential attacks on it by an opposed party. The most important general conclusion to be drawn from these observations is that many of the most important fallacies, and particularly those closely related to the notion of burden of proof, are amenable to precise and useful analysis only if argumentation is viewed as part of a goal-directed dialogical process for arriving at a reasoned conclusion by weighing the relevant arguments for and against it. These considerations take us to the conclusion that the evaluation of an argument in a given case as fallacious or not depends on the assumption that the argument was being used for some conversational purpose in a dialogue between two parties.

As shown in Chapter 1, the use of defeasible arguments that might turn out to be fallacious in some instances is manageable provided the normative model of argumentation is that of a dialogue structure in which evidence continually comes in during the exchange of arguments in the argumentation stage of the dialogue. The problem is that the participants have to be open-minded, because all arguments of the defeasible kind need to be regarded as falsifiable in principle. There has to be a knowledge base set in place at the opening stage representing a database of propositions that the participants can agree is based on common knowledge. Moreover, this knowledge base has to be treated like a commitment set that is open to additions and retractions as the dialogue proceeds.

5. Dialogue Systems and Fallacies

As shown in Chapter 1, a dialogue structure has a goal, and it starts from an opening stage, moves through an argumentation stage and concludes in the closing stage. A framework of evaluation is presupposed (Walton, 1995, 261) in which the actual text of discourse surrounding the argument needs to be taken into account and modeled as a dialogue that is moving forward from an opening stage to the supposed fulfillment of a goal at the closing stage. The context dependence of fallacy identification and evaluation is shown to be context dependent by van Eemeren (2010). Not only does the text of discourse of the actual discussion need to be taken into account, but also the projected completion of the dialogue as it moves toward its goal.

In a dialogue structure of a kind outlined in Chapter 1 the two participants, called the proponent and the respondent, take turns making moves. Each has what is called a commitment set (Hamblin, 1970; 1971). A commitment set, in the simplest case, is just a set of statements. They could be written on a blackboard, for example, so that both commitment sets are visible to both participants. The rules in a dialogue system govern which kinds of moves can be made, how the other party must respond to a given move at the next move and what happens to this commitment set at each move. Commitment rules determine which statements go into or are taken out of each commitment set at each move (Prakken, 2000). For example, if the proponent asserts statement A at some move, then A is added to his or her commitment set. If a participant retracts commitment to statement B at some point in a dialogue, and he or she was committed to B previously, then B is now removed from his or her commitment set. Among the most difficult problems is the formulation of rules of retraction of commitments for various types of dialogue (Krabbe, 2001).

Commitment rules for several types of dialogue are set out in Walton and Krabbe (1995). One of the most common types of dialogue is called persuasion dialogue. 'Persuasion' in formal dialogue theory refers not to psychological persuasion but to rational persuasion (Prakken, 2006). A proponent persuades a respondent to commit to a statement A in this sense by presenting a structurally correct argument with A as the conclusion containing only premises that are already commitments of the respondent. The goal of persuasion dialogue has been accomplished in such a case because the respondent was not committed to this particular statement, but now he or she is. Several different formal models of persuasion dialogue have been constructed in Walton and Krabbe (1995). Some, called rigorous persuasion dialogues (RPDs), have rules that do not allow the participants much flexibility, but the advantage of an RPD is that it is fairly simple while at the same time being formally rigorous. But RPDs model argumentation only in a simplistic way that does not express many of the important features of empirical cases of natural language argumentation. Another kind, called permissive persuasion dialogue (PPD), is more flexible and is closer to empirical argumentation.

No matter which type of dialogue is concerned, and no matter which rules are selected – and there can be many variations – arguments are always evaluated in light of three factors. These are how the argument was put forward in a dialogue, how that move affects the commitment sets of both parties and, in some cases, how the respondent replied to the argument. The device of commitment is useful and does not get into all the problems that have been encountered with the BDI model of defeasible reasoning as belief revision. A belief, as explained in Chapter 1, is an internal entity, and using the BDI model can involve an argument evaluator in the mysterious metaphysics of iterated beliefs. Commitment is a less opaque concept. You are committed only to statements you have gone on record as accepting in a dialogue. The idea is that a public record is kept of a participant's set, as each move is made, so that if an arguer claims that he or she never said something, the other party can go back over what he or she said or did not say and use this as evidence in determining commitment. Thus commitment represents acceptance of a specific statement, judged by the evidence available from the prior text of dialogue in a given case. Dialogue models represent argumentation in a dynamic way. An argument is not just a static set of premises and conclusion, but is a speech act put forward by one party and replied to by the other party. This capability to represent defeasible argumentation is the key asset in enabling dialogue models to assist argumentation schemes in the analysis and evaluation of cases involving fallacies.

As indicated in Chapter 1, there are different types of dialogue. In a persuasion dialogue, the goal is to resolve or throw light on some conflict of opinions or unsettled issue (Prakken, 2006). The proponent of an argument tries to get the respondent to commit to the conclusion by using the premises as reasons. The proponent uses the commitments of the other party as these premises. If he or she puts forward a structurally correct argument that has premises that are all commitments of the respondent, then the respondent is rationally obliged to commit to the conclusion. That is the process called rational persuasion. But there are other types of dialogue as well (Walton and Krabbe, 1995). As shown in Chapter 7, scientific inquiry takes the form of an investigation that collects facts and then tries to prove or disprove some statement using these supposed facts as premises. The rules regarding burden of proof, and other rules that are significant in diagnosing fallacies, vary from one type of dialogue to another. However, all formal models of dialogue of the kind used for studying fallacies have a common structure.

A dialogue, in the simplest case, has two participants, called the proponent and the respondent. The two participants take turns making moves. The moves are essentially speech acts of various kinds. For example, asking a question is a kind of move. Making an assertion is a kind of move. Putting forward an argument is another common kind of move. A type of dialogue is defined formally as a set of participants, a set of rules defining permitted or required moves, a set of rules for determining how one participant must reply to the prior move of the other participant and a set for determining when a completed sequence of moves fulfills the goal of the dialogue (so-called win-loss rules). The general idea is that a dialogue is a sequence of moves, starting at a first move and ending at a last move. In the dialogue theory of Hamblin (1970; 1971), the proponent makes the first move, the respondent makes the next move, and then they take turns, producing an orderly sequence of moves. Each member in the sequence is defined by Hamblin (1971, 130) as a triple, $\langle n,p,l \rangle$, where n represents the length of the dialogue (the number of moves so far) and p is a participant. And l is what Hamblin calls a locution, or what we now call a speech act. Such systems of dialogue have been proposed by Mackenzie (1981; 1990) and Walton and Krabbe (1995).

A dialogue should be seen as having three characteristic stages (Gordon and Walton, 2009). The sequence of argumentation in the argumentation stage should be seen as having started at the opening stage and moving toward the closing stage. The three stages are shown in Figure 8.1.

During the opening stage, the rules for allocating the requirements for what constitutes a winning argument that will apply over the following two stages are set, and both parties become committed to following these rules in order to settle the issue by rational argumentation. Next, there is an argumentation stage in which the arguments and rebuttals on both sides are brought forward and replied to. Each side takes its turn to make moves. Finally, there is a closing stage in which it is judged whether the issue has been settled.

Each type of dialogue has its own special rules of procedure that need to be followed by both sides during the argumentation stage. The question is whether fallacies can be identified simply with breaches of the rules. In some cases they can, and the rules provide valuable normative guidelines that give insight into what is wrong about a fallacious argument or move. However, there is no one-to-one correspondence between a particular fallacy and the violation of some particular procedural rule of a dialogue. As we look over the various fallacies, the problem is that a single fallacy can be committed in a number of ways. One such rule is that a participant in a critical discussion must not prevent the other party from putting forward arguments or asking questions that are legitimate contributions to the dialogue.

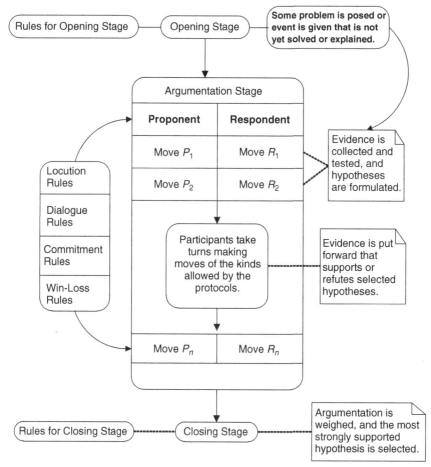

FIGURE 8.1 The Three Stages of a Dialogue

However, several of the traditional major informal fallacies appear to violate this rule. Appeal to force is one, but many of the other major informal fallacies also appear to be fallacies because they commit such an infraction. As shown above, *argumentum ad verecundiam* often fits this diagnosis. And, as will be shown below in Section 8, the *secundum quid* fallacy, the fallacy of ignoring exceptions to a generalization, can be held to be fallacious when the arguer ignores critical questions that should be raised about exceptions to a general rule.

In some cases, fallacious arguments are too weak because a required premise has not been supported adequately by bringing forward enough evidence, but despite this weakness, because the argument is so powerfully impressive to the given audience at a particular moment, it has a devastating impact and carries the day. A good example of this phenomenon is the *ad hominem* type of argument, such as the use of negative campaign tactics at the right moment in a political campaign. Such an argument may be so powerful and convincing to an audience that it is taken for a much stronger argument than it really is. Perhaps the argument is based on very little or no evidence that can be verified and is merely innuendo based on suspicion or some rumor from an unspecified source. Despite these deficiencies, the argument may still be very effective and swing voters one way or the other, especially when the time for refuting it, or even seriously examining it, is not available because the time between the negative attack and the election is very short. In other cases, the problem is not so much the weakness of the argument, its lack of support, but the way it is used as a powerful tactic to close off further discussion.

6. Profiles of Dialogue

Using the whole apparatus of a formal dialogue structure with all its stages and rules (see Section 5 above) may not be necessary to help analyze a text of argumentation in some examples. Often the most useful tool is the profile of dialogue (Krabbe, 1999). A profile of dialogue is a relatively short table of moves with the proponent's moves listed sequentially in the left column and the respondent's matching moves (MV) in the right column. An example is the small profile of dialogue shown in Table 8.1. In this example, the proponent began at move 1 by asking a why-question. The respondent replied to the question by putting forward an argument, giving a reason why the proponent should accept the statement *A* that he or she questioned.

As the dialogue proceeds, the respondent keeps trying to persuade the proponent to accept *A*. He or she uses a *modus ponens* form of argumentation at moves 3–5 to try to get the proponent to accept *C*. By this means he or she hopes to get the proponent to accept *B* and, ultimately, *A*.

How the profile of dialogue works as applied to an example of the fallacy of many questions above is shown in the profile presented in Table 8.2 (Walton, 1995, 203), representing the proper sequence of turn-taking by both parties. The question asked is, 'Have you stopped cheating on your income taxes?' The reason that asking this question is taken to be fallacious is that the respondent may never have cheated on his or her income taxes or may never have even filed income tax returns in the past. But if he or she has to answer the question directly, yes or no, he or she cannot deny these assumptions.

Once the questioner has made moves 1 and 2, and the respondent has answered as shown in Table 8.2, then finally the questioner can properly ask the complex question.

A profile of dialogue can also be represented as a graph structure, as shown in the example of the question, 'Are you confused when you are

TABLE 8.1 *Example of a profile of dialogue*

MV	Proponent	Respondent
1.	Why should I accept *A*?	Because *B*
2.	Why should I accept *B*?	Because *C*
3.	I do not accept *C*	Do you accept 'If *D*, then *C*?'
4.	Yes	Do you accept *D*?
5.	Yes	Then you must accept *C*

TABLE 8.2 *Profile of dialogue for the fallacy of many questions*

MV	Questioner	Respondent
1.	Have you made income tax returns in the past?	Yes
2.	Have you cheated on those income tax returns in the past?	Yes
3.	Have you stopped cheating on your income tax?	

not on your medications?' Asking this question could be considered to be fallacious if it is used as an attempt to suggest that the respondent may not be mentally stable. In turn, that conclusion could be used as the basis of an *ad hominem* attack alleging that the person may not be mentally stable and that therefore no serious attention should be paid to his or her argument. A profile of dialogue for this example is shown in Figure 8.2.

Tracking the sequence of questioning shown in the profile in Figure 8.2, we can see that no matter which way the respondent answers, he or she is inevitably led to one or more of the dangerous admissions shown in the darkened boxes that represent dangerous admissions.

The profile in Figure 8.3 shows the proper sequence of questions and answers for a nonfallacious instance of the same complex question.

In Figure 8.3, the following correct sequence of questioning is represented. Before asking the final complex question shown at the bottom of Figure 8.3, the questioner must first ask the two prior sequences of questions shown above it. If all the questions are asked in the right order, as shown in Figure 8.3, asking the complex question in that context of dialogue could be reasonable. This profile illustrates the point that the same complex question can be fallacious or not, depending on the context of dialogue, referring to the sequence of questions and replies and the order in which they were put forward.

This discussion suggests that requirement 2 in the definition of 'fallacy' proposed in Section 1 of Chapter 8 needs to be modified as follows: a fallacy is a failure that occurs in what is supposed to be an argument or an argumentation strategy used as a move in a dialogue. Two other potential exceptions to requirement 2 also need to be considered. Common instances of

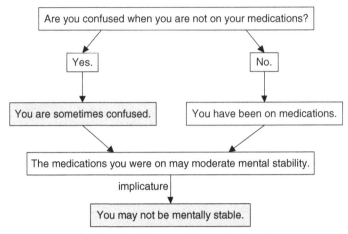

FIGURE 8.2 Profile 1 for the Fallacy of Many Questions

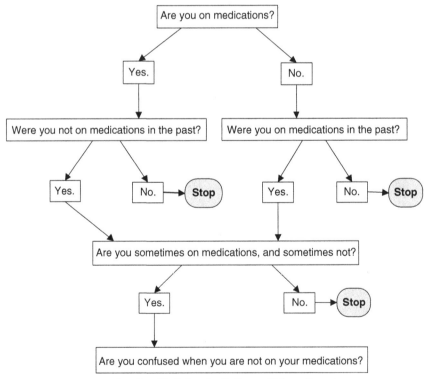

FIGURE 8.3 Profile 2 for the Fallacy of Many Questions

the fallacy of begging the question and the fallacies of relevance, coming under the traditional heading of *ignoratio elenchi*, involve chains of argumentation rather than single arguments. The modified version of requirement 2 makes room for these kinds of fallacies as well. The profile of dialogue represents the ordered moves in a dialogue before and after the move where the argument in question was put forward. The profile of dialogue is a useful tool for the analysis of argumentation associated with fallacies (Krabbe, 1999). A fallacy can also be a sequence of argumentation moves in a dialogue.

7. Dialogue Conditions for Defeasible Arguments

To illustrate how defeasible reasoning can be modeled in a dialogue format, consider an instance of the Tweety argument displayed in the profile in Table 8.3.

In the Tweety profile, the proponent put forward a defeasible argument. At move 3, the respondent conceded the major premise of the argument. At move 4, the proponent drew the conclusion. But then, at the respondent's turn in move 4, the respondent defeated the argument by bringing in new information about Tweety. This profile shows how things should go ideally in a dialogue in which one party has brought forward a defeasible argument. The proponent properly used a defeasible generalization as a premise. This premise, along with the other one, formed an argument having the form of DMP (defeasible *modus ponens*). The respondent indicated at move 2 that the premise that Tweety is a bird is accepted. At move 3, acceptance of the generic premise that birds fly is indicated. Thus at the next move, the respondent must commit to the conclusion that Tweety flies, in order to follow the commitment rules of a persuasion dialogue. The big problem is whether the respondent can now retract commitment from one of the premises. At move 4, the respondent does not retract directly. Instead, the respondent offers a new bit of evidence in the form of the statement 'Tweety is a penguin'.

The Tweety profile of dialogue illustrates a key feature of defeasible argumentation. The proponent conceded that Tweety does not fly, and so his prior argument is now defeated. An important feature of a defeasible argument in a dialogue is the *Openness to Defeat (OTD) Condition*: When the proponent has put forward a defeasible argument during the argumentation stage in a dialogue, he or she must be open to giving it up and admitting its defeat at any future move by the respondent that defeats the argument at any point before the closing stage of the dialogue. Principles somewhat like the OTD condition have been recognized in the argumentation literature. Van Eemeren and Grootendorst (1992, 108) state that parties in a critical discussion cannot declare their viewpoint as sacrosanct, so that they are rendered immune to criticism. Johnson (2000, 224) formulated a

TABLE 8.3   *The tweety profile of dialogue*

MV	Proponent	Respondent
1.	Tweety flies	How can you prove that?
2.	Tweety is a bird	So what?
3.	Birds fly	Yes, they generally do
4.	Therefore, Tweety flies	But Tweety is a penguin
5.	Oh, really?	Therefore, Tweety does not fly

principle of vulnerability of arguments comparable to Popper's notion of falsifiability: "to be a legitimate argument, an argument must be vulnerable to criticism".

What sort of move defeats a defeasible argument such as the Tweety argument? The defeater has to be an exception to the rule postulated by the generic premise. Such a defeasible rule is subject to exceptions. So the respondent, to defeat the argument, has to come up with information in the given case that presents an exception to the rule. The OTD condition requires that if the respondent comes up with such a case, and the proponent admits that it constitutes an exception to the rule, then the proponent has to give up his or her argument. Thus the respondent no longer has to accept the conclusion. Indeed, both parties must now retract commitment to the conclusion.

The OTD condition says something about all future moves of a dialogue between the move where the defeasible argument was put forward and the closing move of the dialogue. It says that at all such moves, the proponent must retract the conclusion, if the respondent makes the sort of move that defeats the argument used to prove it. There is no way to predict when a respondent may come up with such a defeater before closure of a dialogue. Defeasible arguments are characteristically open-ended.[7] They have to do not just with the argument as presented, but with the future stretch of dialogue beyond the argument in a continuing dialogue that provides the framework of argument use.

A second feature of defeasible argumentation illustrated by the Tweety dialogue is the phenomenon often called the shifting of the burden of proof. At move 1, the proponent has made a claim, and so has incurred a burden of proof. The respondent asks, "How can you prove that?" – referring to the proponent's claim that Tweety flies. The proponent then fulfills this burden by putting forward a defeasible argument to support it. That should be the end of it, if the argument was not defeasible. But then at

---

[7] This observation presents a computational problem. For this reason, much of the modeling of dialogues in computing uses the restriction that an argument must be attacked immediately or conceded. See, for example, Vreeswijk and Prakken (2000).

TABLE 8.4 *An example of a profile of dialogue fitting the DSD*

MV	Proponent	Respondent
1.	P, D; therefore C	I accept that argument
2.	Do you accept premise P?	Yes
3.	So you must accept C	Yes
4.	C has been proved then?	I do not accept C
5.	Why not?	Here is evidence for not-D
6.	Okay, I accept that evidence	Will you now retract C?

move 4, the respondent asserts that Tweety is a penguin. The respondent made an allowable move by bringing forward an exception to the rule posited by the proponent's major premise in his or her prior argument. This pattern is characteristic of defeasible argumentation.

The pattern of argumentation characteristic of defeasible argumentation has the following characteristic called the dialogue sequence for defeasible argumentation (DSD).

1. The proponent puts the defeasible argument forward.
2. One premise is a generalization or conditional (rule) that admits of exceptions.
3. If the respondent is committed to the premises, he or she must commit to the conclusion.
4. But his or her commitment to the conclusion can be retracted.
5. The dialogue must remain open to the respondent's finding an exception to the rule.
6. As soon as the respondent cites such an exception, the proponent's argument defaults.
7. The proponent must now retract commitment to the conclusion.

The argumentation in the profile of the dialogue shown in Table 8.4 fits the DSD pattern.

Let's say that the dialogue is a critical discussion and that C is a statement that the proponent needs to prove, in order to prove his or her ultimate thesis in the dialogue. Let's also assume, however, that the respondent can still make a reply at move 4. The closing stage has not been reached yet. If the respondent can find an exception to the default rule in the proponent's defeasible argument put forward at move 1, the proponent must retract C. Normatively speaking, the latter must retract C because the DSD applies. Hence the proponent must answer 'yes' at the next move. If the answer is 'no', the proponent commits the *secundum quid* fallacy.

This sequence of argumentation has the reversal of burden of proof characteristic of the typical presumptive argument of the kind often used in law. A presumption is accepted tentatively in a dialogue as a way of moving the dialogue forward, but it can be defeated if new evidence comes into the dialogue. For example, in law, a person who has disappeared without a trace for a determined number of years may be presumed dead, for purposes of settling an estate. The basis of the reasoning is an argument from ignorance. But in this instance, the argument from ignorance is reasonable, not fallacious, as long as it is treated as a defeasible argument that can default in the face of new evidence. If there is no evidence that the person is alive, then the conclusion can be drawn by inference (after the stated period) that he or she is dead. But if such a person turns up, the presumption that he or she is dead is defeated. This type of argumentation will be treated at more length in Section 8, concerning arguments from ignorance. If it turns out later that the person who turned up is not the missing person, then the new conclusion will have to be retracted and we may revert to the old one. The theory is that we track defeasible chains of argumentation through a dialogue.

Retraction of the conclusion of the argument is shown as a key dialectical feature of defeasible argumentation at stage 7 of the DSD. In a persuasion dialogue, retraction is allowed, although not in all circumstances. Retraction of one statement may require retraction of other statements to which it is closely related by inference. For example, if the proposition that a person is dead is retracted, then the proposition that someone else may inherit his or her estate may also need to be retracted, even though it was accepted previously. However, the problem of retraction has not been solved for persuasion dialogue, even though proposed solutions to it have been put forward (Krabbe, 2001). Thus in the dialogical theory of defeasible reasoning, the problem of defeasibility is recast as a problem of determining conditions for retraction in the various types of dialogue.

8. The Fallacy of *Secundum Quid*

The single fallacy most closely related to defeasible reasoning in general is the one called *secundum quid*. The *secundum quid* fallacy was very clearly explained in one logic textbook (Joseph, 1916, 589), in a way that makes it appear similar to the way Aristotle described it. However, as shown in detail in Walton (1999b), its treatment in nearly all the other logic textbooks is mixed in with other fallacies and defined using ancient and unhelpful terminology, such as "accident" and "converse accident". The textbook treatments of this fallacy and other related errors of reasoning are more than just inconsistent. They are highly confusing. *Secundum quid* is mixed up with related fallacies such as hasty generalization, glittering generality, converse accident, oversimplification and *de dicto simpliciter*. These supposed fallacies are conflictingly defined and mixed in together in a way that shows an urgent need for clarification and systematization. The best place to begin is with some examples.

Secundum quid (in Greek, *para to pe*), means 'in a certain respect' and refers to qualifications attaching to a term or generalization (Hamblin, 1970, 28). The *secundum quid* fallacy is the error of neglecting qualifications when drawing a conclusion by inference. The clearest account of what this fallacy is supposed to consist in was given by Joseph (1916, 589), using this example: "Water boils at a temperature of 212 degrees Fahrenheit; therefore boiling water will be hot enough to cook an egg hard in five minutes: but if we argue thus at an altitude of 5,000 feet we shall be disappointed; for the height, through the difference in the pressure of the air, qualifies the truth of our general principle". The fallacy of *secundum quid* cited here can be specified more precisely by examining the structure of the following argument.

Explicit General Premise: Water boils at a temperature of 212 degrees.

Implicit General Premise: Immersing an egg in water at a temperature of 212 degrees Fahrenheit will be hot enough to cook an egg hard in five minutes.

Conclusion: Immersing an egg in boiling water will be hot enough to cook an egg hard in five minutes.

This argument works fine, assuming that standard conditions, such as those of altitude, hold. But let's add a new premise, the statement that we are at 5,000 feet, a higher altitude than we might be assuming to be standard. What happens is that the explicit general premise fails to hold, not generally, but in these specific conditions. In these conditions water boils at a different temperature, and the inference defaults. The error here can be diagnosed as one of neglecting qualifications to the explicit general premise of the argument, and thereby drawing a wrong conclusion.

The fallacy here seems to be precisely an error of defeasible reasoning. The initial argument was fine, as applied to what is taken to be a normal or standard situation, but defaults when new knowledge comes in about the particulars of the case. Once the new data comes in, the old conclusion must be given up. It no longer holds in the special circumstances of the case. It's not that the old argument was bad, generally speaking. It's just that it doesn't work, once the new data about altitude is added in as a premise.

According to Aristotle's description of *secundum quid* (*On Sophistical Refutations* 167a10-167a13), this fallacy arises from two ways an expression can be used: absolutely or in a certain respect. "This kind of thing is easily seen by anyone, e.g. suppose a man were to secure the statement that the Ethiopian is black, and were then to ask whether he is white in respect of his teeth; and then, if he be white in that respect, were to suppose at the conclusion of his questions that therefore he had proved dialectically that he was both white and not white (*De Sophisticis Elenchis* 167a10–167a13; Ross, 1928). The premise that the Ethiopian is black generally (meaning,

TABLE 8.5 *Profile of dialogue for the ethiopian example*

MV	Proponent	Respondent
1.	Is the Ethiopian black?	Yes
2.	Is he white in respect to his teeth?	Yes
3.	If something is white, must it follow that it is not black?	Well, yes, of course
4.	So by your admission, the Ethiopian is black and not black	Respondent loses

in respect of skin color) is true. But from this premise it would be errone-ous to draw the conclusion that he has black teeth. For when we say 'The Ethiopian is black', we (normally) mean that he is black in a certain general respect, one that does not refer specifically to the color of his teeth.

The Ethiopian example is not quite the same as the egg example, because the former refers to a generality implicit in standard word usage while the lat-ter refers to an empirical generalization. Also, the analysis of the egg example based on defeasible reasoning does not seem to apply to the Ethiopian exam-ple. We now need to recall that in the section on defeasible reasoning above (Section 4), following Hart a distinction was drawn between defeasible con-cepts and defeasible reasoning in arguments. It is this distinction that brings out the difference between the defeasibility in the two examples. The egg example is a standard case of defeasible reasoning of the kind usually taken to be central in computing, corresponding to the description of defeasible reasoning cited in Hart's chapter in Section 4. The Ethiopian example is an instance of a defeasible concept. When we say the Ethiopian is black, argu-ing that he cannot be black because his teeth are white is fallacious because that is not an exception. As we saw, Aristotle offered a dialectical analysis of the Ethiopian example, showing how failing to draw the distinction between 'absolutely' and 'in a certain respect' leads to a contradiction.

This analysis can be displayed using the profile in Table 8.5 (Walton, 1999a, 164).

The profile in Table 8.5 shows that while each of the respondent's single replies answers the proponent's question reasonably, the ordered dialecti-cal sequence of them leads to inconsistent commitments. In Aristotelian dialectic, a fallacy can be analyzed as a set of opinions that seem individually plausible (endoxic, generally accepted) but that are collectively shown to lead to a contradiction (or to a comparable logical difficulty). The analy-sis reveals that one should be careful in accepting a statement to be clear whether the statement is meant to be true absolutely, in an unrestricted sense, or whether it is meant to be true only in a certain respect.

This is not the end of the story on the fallacy of *secundum quid*. How general rules should be applied to particular cases is centrally important in scientific reasoning as well as legal reasoning of the most common sort. There are special argumentation schemes (Walton, Reed and Macagno, 2008, 343–345) representing arguing from rules and arguing from exceptions to rules. Also, more needs to be done to map the relationship of *secundum quid* to related fallacies such as hasty generalization.

Both the egg and the Ethiopian example show how fallacies can arise when an arguer is too rigid, and views an inference as based on a strict universal generalization modeled by the universal quantifier in deductive logic instead of a defeasible generalization. The fallacy can be seen as a species of violation of the OTD condition. It can be argued that many of the fallacies, especially the twelve based on argumentation schemes, are failures of this sort (even though other faults can be involved as well). Let us next consider another fallacy that is also closely associated with defeasible reasoning.

9. Lack of Knowledge Inferences

The simplest formulation of the scheme for the *argumentum ad ignorantiam* of the logic textbooks is this: statement A is not known to be true (false), therefore A is false (true). As noted above, this form of argument is often called the lack of evidence argument in the social sciences or an *ex silentio* argument in history, where it is presumed not to be fallacious. In both fields it is commonly regarded as a reasonable but inconclusive form of argument. To cite an example (Walton, 1999c), there is no evidence that Roman soldiers received posthumous decorations, or medals for distinguished service, as we would call them. We have evidence only of living soldiers receiving such awards. From this lack of evidence, it has been considered reasonable by historians to put forward the hypothesis that Roman soldiers did not receive posthumous decorations. Of course, such a conjecture is based not on positive evidence, but only on a failure to find evidence that would refute it. Such arguments from ignorance are common in many fields, not least in law, as will be shown below.

As shown in Walton (1999a), argument from ignorance needs to be seen as an inherently dialectical form of argumentation. The context most often helpful to grasping the structure of this form of argumentation is that of an ongoing investigation in which facts are being collected and inserted into a knowledge base. In such a context, the argument from ignorance can be represented using the following argumentation scheme for epistemic argument from ignorance given in Walton, Reed and Macagno (2008, 328). In this knowledge-based scheme, D is a domain of knowledge and K is a knowledge base in a given domain, or field of knowledge.

All the true propositions in D are contained in K.
A is in D.
A is not in K.

For all *A* in *D*, *A* is either true or false.
Therefore, *A* is false.

This form of argument can be deductively valid in a domain *D* where *K* is closed, meaning that it contains all the statements that can ever be known in that domain. But in a vast majority of cases, argument from ignorance is a defeasible inference that may default as an investigation proceeds and new knowledge comes in. Thus one of the most important critical questions in evaluating any given instance of an argument from ignorance is whether the knowledge base is open or closed.

An example that can be used to illustrate how the nonfallacious *ad ignorantiam* works as an argument is the foreign spy argument: Mr. X has never been found guilty of breaches of security or of any connection with agents of the foreign country for which he is supposedly spying, even though the Security Service has checked his record; therefore, Mr. X is not a foreign spy. This argument from ignorance is defeasible, because it is not possible to be absolutely certain that Mr. X is not a foreign spy. Mr. X could have avoided detection through many security searches, as Kim Philby did. Hence arguments from ignorance tend to be defeasible arguments, even though they can be conclusive in some cases. The argument from ignorance can be seen to be a very common kind of defeasible argumentation, once you learn to recognize it.

There is a very common principle often appealed to in knowledge-based systems in artificial intelligence called the closed world assumption (Clark, 1978). Essentially, the closed world assumption means that all the information that there is to know or find is listed in the collection of information one already has, but there are different ways of representing information. According to Reiter (1980, 69), the closed world assumption is met if all the positive information in a database is listed, and therefore negative information is represented by default. Reiter (1987, 150) offers the example of a database for an airline flight schedule to show why negative information is useful. It would be too much information to include in such a database all flights and the city pairs they do not connect. This amount of information would be overwhelming. Instead, the closed world assumption is invoked. If a positive flight connection between a pair of cities is not asserted on the screen representing the database, the conclusion is drawn that there is no flight connecting these two cities. If the system searches for a flight of the designated type and does not find one in the database, it will reply "no". Reiter (1980, 69) described the form of argument used in this sequence of knowledge-based reasoning as: "Failure to find a proof has sanctioned an inference." As noted above, this kind of inference by default from lack of knowledge has traditionally been called the *argumentum ad ignorantiam* in logic. The argument from negative evidence may be merely a defeasible inference that leads to a provisional commitment to a course of action, but should be seen as open to new evidence that might come into an investigation and needs to be added to the knowledge base.

In legal argumentation, argument from ignorance is closely associated with what is often called the presumption of innocence in a criminal trial.[8] The prosecution has the burden of proof and must bring forward enough evidence to satisfy the proof standard of beyond reasonable doubt. The defendant need only bring forward enough evidence to prevent the prosecution from meeting its burden of proof, by casting enough doubt on the prosecution's attempt to prove its claim. This asymmetry involves an argument from ignorance. If the defense can show that there is a lack of evidence to support the prosecution's claim (ultimate thesis to be proved in the trial), then the defense has shown that this claim does not hold up and must be rejected. This form of argumentation meets the requirements for the argumentation scheme of the argument from ignorance. Thus argument from ignorance is fundamental to the argumentation structure of the trial in the adversary system.

Argument from ignorance is also common in more special forms in legal argumentation. For example, as shown by Park, Leonard and Goldberg (1998, 103), there is a presumption that some writing has been accurately dated: "unless the presumption is rebutted, the writing in question will be deemed accurately dated". Another example (153) concerns character evidence. Suppose a first person was in a position to hear derogatory statements about a second person if any were made. And suppose the first person testifies that he or she heard no such comments. This testimony counts as evidence of the first person's good character. The form of argument in such a case is that of argument from ignorance. If no evidence of bad character was found or reported by the witness, this lack of such a finding may be taken as evidence of good character.

Arguments from ignorance having the form of the argumentation scheme set out above are best analyzed as defeasible arguments at some stage of a dialogue or investigation in which evidence is being collected and assessed. The typical pattern of reversal of burden of proof characteristic of the *argumentum ad ignorantiam* can be modeled by the following dialogue sequence, discussed by Krabbe (1995, 256).

Proponent: Why *A*?

Respondent: Why not-*A*?

This pattern of shifting of the burden of proof in dialogue is characteristic of the argument from ignorance. In a case study in Walton (1996c, 118–122),

---

[8] This terminology is misleading. It should be said that the burden of persuasion is on the prosecution side.

during a political debate the opposition party asked the government minister to prove with absolute certainty that Canadian uranium was not being used for military purposes. The minister replied that the opposition should give evidence to support their allegation that Canadian uranium was being used for military purposes.

Argument from ignorance is not always a defeasible argument. In a case where a knowledge base is closed, the argument from ignorance can be conclusive, assuming that the closed world assumption holds, for example, if a dialogue has reached the closing stage. Thus, generally speaking, arguments from ignorance need to be analyzed and evaluated using two tools. One is the argumentation scheme. The other is the placement of that scheme in a context of dialogue representing an investigation that has some standard and burden of proof set at the opening stage, so that it is known whether the closed world assumption applies or not.

Argument from ignorance displays a pattern common to many if not all of the twelve fallacies based on schemes. That pattern is for the proponent to press ahead too aggressively to jump to a conclusion uncritically by overlooking the defeasibility of the argumentation scheme in question. The proponent may have collected some evidence, even in the form of negative evidence, but has not collected enough to satisfy the standard of proof that should be required. When confronted by critical questions, or by evidence that suggests an exception to the defeasible rule on which the inference is based, the proponent violates the OTD condition. There are various ways to try to carry out such a strategy by trying to escape the need for retraction. Each way is associated with a single fallacy or a group of fallacies. One way is to try to browbeat the respondent into accepting the conclusion (e.g., by arguing that nothing he or she says is of any value, because he or she is not an expert). Another is to try to shift the burden of proof to the other side to avoid proof (as in argument from ignorance). Still another is to mount a personal attack (*ad hominem* argument) that attempts to discredit the respondent, so that any objections or counterarguments he or she raises are discounted as worthless.

10. Conclusions

The dialogue-based theory applied in this chapter has identified three characteristics that are essential to the defeasible argumentation on which the twelve scheme-based fallacies are built: openness to defeat, reversal of burden of proof and retraction of commitment. It was shown how these three characteristics can best be modeled using a dialogue model in which an argument is seen as a sequence of moves made by a proponent and a respondent who take turns. With respect to defeasibility, there are two ways to view a dialogue in a given case. It can be open or closed. In typical cases, defeasible arguments are viewed as having been put forward, and as not yet being defeated, assuming that the dialogue is still open. At the opening or argumentation stages, commitment to the conclusion of the argument is still tentative and subject to possible retraction as the dialogue proceeds. This tentativeness has often been the basis of a traditional feeling of distrust about defeasible arguments. They have often been seen as unreliable or even fallacious. The argument from ignorance is an excellent case in point. We need to move beyond this wholesale rejection of defeasible arguments and see them as arguments that can go wrong but are often quite reasonable. They are typically reasonable in cases of uncertainty and lack of knowledge where some decision for action needs to be made or a presumption adopted. They are dangerous but necessary.

One of the most important things about defeasible arguments is that they are often used during the argumentation stage of a dialogue, before it has reached the closing stage. One needs to take a stance of being open-minded about such arguments and to be ready to give them up if new evidence comes in. It is perhaps because of the human tendency to be reluctant to admit defeat, and thus to close off argumentation too soon, that the problems with defeasible arguments often arise as fallacies. These matters have to do with dialectical factors such as presumption, default, burden of proof and openness and closure of argumentation in a dialogue. As shown in this chapter, such arguments depend on a default blocker that makes acceptance of an argument conditional on lack of evidence to support a key assumption. Such an argument is based on a mixture of knowledge and lack of knowledge. It is very often an argument from ignorance.

In an argument from ignorance, it is argued that a statement *A* is not yet known to be true (false), and argued on this basis that *A* is false (true). The premise may be taken to mean that *A* is not yet known to be true (false) at some point in an investigation or collection of data that is under way. At least, that is what the premise means if the case is a typical sort where the dialogue is not closed, and the argumentation is still open to possible defeat in the future as more information is collected. The proponent urges the respondent to accept the conclusion that *A* is false (true) based on the findings to that point in the dialogue. If the respondent does accept the conclusion of the argument from ignorance, it is a tentative commitment that is open to retraction as the dialogue proceeds. Only when the dialogue reaches the closing stage does the argument from ignorance become non-defeasible. The conclusion is then based on the closed world assumption.

The notion of fallacy should be defined by a more concise but amplified definition that improves on the pragmatic definition (Walton, 1995) proposed in Section 1.

- An argument
- that is often an instance of a defeasible argumentation scheme

- that is reasonable, but is somehow used wrongly, and
- that falls short of the standard of proof set for it in the dialogue the arguer is supposed to be taking part in
- but that plausibly seems correct [in its given context of dialogue]
- and committing it poses a serious obstacle to reaching the goal of the dialogue.

The tools shown to be most useful for the study of fallacies in this chapter are formal models of dialogue, schemes with matching sets of critical questions, defeasible reasoning, openness to contra evidence in the argumentation stage of a dialogue, burden of proof, strategic maneuvering, profiles of dialogue, the OTD condition and the DSD sequence.

9

The Straw Man Fallacy

Straw man is a modern addition to the list of informal fallacies treated in logic textbooks, where it is said to be the fallacy of misrepresenting an opponent's real position as a weaker one that can more easily be attacked (Johnson and Blair, 1983; Freeman, 1988; Govier, 1992; Walton, 1996c). However, much earlier, Aristotle remarked (*Topics* 159b30–35) on the danger in argumentation of misrepresenting another arguer's position, and no doubt philosophers have often made similar comments in the past. Thus although straw man represents the extreme kind of case of misrepresenting another arguer's position, it would be useful more generally to have some tool that could assist in making objective determinations of when an expressed position has been wrongly represented in a given case. For it is a rule of rational argumentation that before you criticize or refute another party's view in a critical discussion, you should be sure that the view you are attacking actually represents the other party's position.[1] This chapter develops a formal dialogue system that can utilize three different kinds of commitment query inference engines that are designed to help in analyzing and evaluating cases where the straw man fallacy is alleged to have been committed.

Such an engine is meant to be used to search through an arguer's commitment set in a case of argumentation that can be structured in a formal dialogue format, so that it can be fairly judged whether a queried statement is in an arguer's commitment store or not. Only if it is not in the arguer's commitment store can an attacker of the alleged position of the second party be fairly judged to have committed the straw man fallacy, based on the evidence in the case at issue. The project undertaken is not to implement any of these inference engines computationally, but only to construct

---

[1] It is a rule of a critical discussion that an attack on a standpoint must relate to the standpoint that has really been advanced by its proponent (van Eemeren and Grootendorst, 1987, 286).

formal models specifying generally how they should work. It is argued that the models are developed to a state of refinement where they can be used as guidelines to assist in dealing with problematic cases in which the straw man fallacy has allegedly been committed.

By using two examples of argumentative exchanges on the popular online discussion forum accessible through Google Groups, Lewinski (2011) has shown that the ability to grasp the straw man fallacy requires a dialectical approach. In both these fragments of online discussions, one arguer attacks another by distorting one of his previous messages in the exchange, and the other party then attempts to denounce and correct the attempted straw man argument. The interactivity of such attacks that are immediately denounced shows that the straw man fallacy requires a dialogue model for its analysis. One cannot merely analyze the argument as an isolated set of premises and conclusion. Instead, a sequence of moves in a dialogue between two parties needs to be taken into account. This means that an adequate analysis of the straw man fallacy in any given case needs to take the context of dialogue into account. The straw man fallacy essentially involves a comparison between one move by one party in a dialogue and a previous move by the other party where commitment was incurred by the other party.

Analyses of some of the examples provided show how some existing tools, such as argument diagramming, are helpful up to a point, enabling a charge of fallacy to be fairly well justified in some instances. The analyses also show that more resources are required to pin down the fallacy by an objective method that has a clear and precise structure. It is argued that we have to go beyond a narrow propositional analysis to apply the commitment query inference engines in a way that takes contextual factors into account. What you have to do, it is shown, is to analyze the discourse to fairly determine what the other party's explicit and implicit commitments are in a rule-governed dialogue the two of you are supposedly engaged in (Hamblin, 1970; Walton and Krabbe, 1995). But how do you carry out this task by some objective method? It is shown that to fully analyze hard cases of straw man arguments, commitment query inference engines need to be supplemented by six pragmatic components embedded in a dialogue framework containing conversational rules for the determination of commitment. The study of these cases helps to expose the complexities of the straw man fallacy in such cases and exposes some limitations of the commitment query inference engines when applied to them.

1. What Is the Straw Man Fallacy?

The straw man fallacy is committed in an interpersonal exchange of arguments when one party describes the position of the other in an inaccurate, even distorted and exaggerated way that makes it appear more unreasonable than his or her real position, thus making it easier for him or her to refute the argument. Consider the example of a debate on cutting down trees in the forest by logging corporations. A moderate environmentalist who was engaged in debate with a representative of the logging industry was confronted with the following attack: "You want to make this province a heaven on earth, but my objection is that it would leave no room for private property and cause massive unemployment!" Presuming that the environmentalist did not make statements of this extreme sort in her previous pronouncements on the subject, and that her position was more moderate than the extreme one portrayed by her opponent, we can say that the attacker committed the straw man fallacy. Hence the straw man fallacy is rightly seen as a fallacy of misattribution of commitment. Whether it has been committed or not in a given case depends on the opponent's commitments in the dialogue, which in turn depend on her prior moves made in the dialogue, as far as these are known in the case.

The contextual nature of the straw man fallacy is brought about by the three necessary conditions for identifying straw man set out by Lewinski (2011, 481–483). First, there has to be a contextualized record of exchanges, meaning that the conversation has to be recorded or at least witnessed by another party. In other words, there has to be some textual evidence that can be used to resolve the allegation that a straw man fallacy has been committed. The second condition is that there has to be some possibility for the accused party to respond and correct the distortion alleged to be the basis of the straw man fallacy. Third, there has to be a careful interpretation of the position held by the first party according to the allegation of the second party based on the textual evidence of what was actually said by the first party.

To my knowledge, straw man had not been included in the logic textbooks in the lists of informal fallacies until the wave of modern textbooks began to appear. Now there are a considerable number of them that include straw man as a fallacy (Walton, 1996d). It certainly was not in Aristotle's original list of fallacies. However, it is clear from many remarks in Aristotle's *Topics* that he recognized the principle that in argument it would be inappropriate to interpret as someone's position an opinion that he did not express or is not committed to, in virtue of what he said. For example, in the *Topics* (159b30–159b35), Aristotle wrote that if the respondent in a dialogue is defending someone else's opinion, then he or she must concede or reject each point in the dialogue in accord with (what he or she takes to be) that person's judgment. This remark refers to something other than the straw man argument, because it is a third party's opinion that is referred to. But it indicates that Aristotle recognized that the changing or distorting of someone's stated position in an argument is an illicit kind of move.

Textbook accounts offer a fairly consistent definition of the fallacy. Johnson and Blair (1983, 71) state that the straw man fallacy is committed "when you misrepresent your opponent's position, attribute to that person a point of view with a set-up implausibility that you can easily demolish, and then proceed to argue against the set-up version as though it were your opponent's." Govier (1992, 157) wrote that the straw man fallacy is committed "when a person misrepresents an argument, theory, or claim, and then, on the basis of that misrepresentation, claims to have refuted the position that he has misinterpreted." According to the definition offered by Hurley (2003, 122), the straw man fallacy is committed "when an arguer distorts an opponent's argument for the purpose of more easily attacking it, demolishes the distorted argument, and then concludes that the opponent's real argument has been demolished".

As noted above, the straw man fallacy can involve exaggeration as well as other forms of distortion of an opponent's position. The following example of a dialogue (Freeman, 1988, 88) illustrates this point. Let's call it the beer and wine example.

Concerned Citizen: It would be a good idea to ban advertising beer and wine on radio and television. These ads encourage teenagers to drink, often with disastrous consequences.

Alcohol Industry Representative: You cannot get people to give up drinking; they've been doing it for thousands of years.

As Freeman (88) points out, the concerned citizen did not maintain that it would be a good idea if teenagers or people in general gave up drinking, and there is no evidence at all that abstinence is the conclusion for which he or she is arguing. However, there is evidence that the alcohol industry representative wants to make us believe that the concerned citizen advocates total abstinence.

What is the straw man fallacy illustrated here, and how can it be proved that it is committed? Freeman offers the evidence needed to answer these questions by stating and contrasting the following pair of propositions (88), and asking which is the easier one to refute.

A: It would be a good idea to ban advertising beer and wine on radio and television (the concerned citizen's original conclusion).

B: It would be a good idea to get people to stop drinking (the alcohol industry representative's portrayal of that conclusion).

Anyone can see that B is much easier to argue against than A. Thus the alcohol industry representative has not only misrepresented the concerned citizen's position but has done so "in a way making it easy to refute, making it look almost silly" (Freeman, 1988, 88). He then attacks the misrepresentation. This case is an excellent one to use as an illustrative example of the straw man fallacy in a textbook. It is an easy case because it is clear that the argument is fallacious and because Freeman's analysis, by exhibiting the contrast between propositions A and B, offers the main evidence needed to prove that the argument commits the straw man fallacy as defined.

Johnson and Blair (1983, 71), as indicated above, defined the straw man fallacy as being committed when the following three conditions are met.

1. The proponent attributes to the respondent a certain view or position.
2. The respondent's real position is not the attributed one, but a different one.
3. The proponent criticizes the attributed position as though it were the one actually held by the respondent.

In this account of the fallacy, the term 'position' may be taken to refer to the commitment set of the respondent. His position is defined by the totality of the commitments that can rightly be attributed to him at any particular point in the progress of a dialogue (Hamblin, 1970; Walton and Krabbe, 1995). These accounts of the straw man fallacy make it clear that its structure has three important components. The first component is the pair of arguers in a dialogue. The second component is the argumentation in which the two are engaged. The third component is the position or viewpoint that each arguer has taken. In this framework it is required for the straw man fallacy to be committed that each party has to base his or her arguments on the position or view the other party has adopted or taken up.

Forms of fallacy closely related to the straw man fallacy, or perhaps even included in it, are refutations based on misquotation and wrenching from context (Walton and Macagno, 2010). In many cases a respondent's quoted words have been changed or otherwise misinterpreted to make his or her position appear other than it really is (as far as can be inferred from what was really said). When this changed version is used against him or her in an argument, the fallacy committed would be a variant on the straw man fallacy. In other cases, the respondent may have been quoted accurately, but what he or she said is placed in a context different from the original one. Then the new context of dialogue is used to draw conclusions that one would not be justified in drawing from the remark in its original context of use. Such cases are on the borderline of the straw man fallacy and related to fallacies of misquotation and wrenching from context.

The following example is taken from related work on the fallacy of wrenching from context (Walton, Reed and Macagno, 2008). It concerns media reporting of Vice President Al Gore that led to the widely circulated story that he claimed that he "invented" the Internet. Gore did not claim he "invented" the Internet nor did he say anything that could reasonably be interpreted as making this claim. The many derisive put-downs to the effect that Gore said that he "invented" the Internet arose from an interview with Wolf Blitzer on CNN's *Late Edition* program on March 9, 1999. When asked to describe what distinguished him from a challenger for the presidential nomination, Gore replied as quoted below in what we will call the "Gore and the Internet" example.

During my service in the United States Congress, I took the initiative in creating the Internet. I took the initiative in moving forward a whole range of initiatives that have proven to be important to our country's economic growth and environmental protection, improvements in our educational system.

This claim to have taken the "initiative" is vague and even ambiguous. A sympathetic interpretation is that he was claiming to be responsible for helping to create the environment (in an economic and legislative sense) that fostered the development of the Internet. However, looking at comparable examples suggest otherwise. Consider this example: "In 1902 President Roosevelt took the initiative in opening the international Court of Arbitration at The Hague, which, though founded in 1899, had not been called upon by any power in its first three years of existence". The language of this example implies that Roosevelt did indeed open the Hague Court in 1902. Or consider another example: "At the 30th U.N. General Assembly, Japan reiterated its intention of continuing to lend its cooperation to the U.N. University headquartered in Tokyo in view of the fact that the University had already begun to operate, and took the initiative in submitting a resolution which sought the positive support of all U.N. member nations for the U.N. University." The language implies that Japan truly submitted a resolution. By comparison, the conclusion is implied in the Gore case that that he claimed, at least in some sense, to have "created" the Internet. Whether this entails that he "invented" it is different, but it comes close. We could rule out this claim for pragmatic reasons related to the principle of charity, like "Gore is an intelligent person and could not have said something that is patently false", and not for any semantic or textual reasons. But if we have independent reasons to be less benevolent toward Gore, we might have a point in accusing him of claiming something false while using an ambiguous phrasing that allows an easy retreat. The Gore example is difficult because the precise wording is not sufficient to sort out the case, and general pragmatic presumptions, themselves defeasible, have to be employed.

At any rate, the uncharitable version was repeated on the media and used by Gore's political opponents to discredit his political views by making him appear to be ridiculous. The problem shown by this example, and that could also be illustrated by many comparable cases of media reporting of political argumentation, is partly one of ambiguity but also one of wrenching from context and one of failure to quote the exact wording of the claim.[2] Such examples show that straw man as a device is a powerful and common argumentation tactic. In this chapter, I concentrate mainly on some central paradigm examples, in order to build up a basic normative structure that

---

[2] A series of textbook examples of the straw man fallacies presented by Carey (2000, 144–146) illustrates these combinations very well. They involve misquotation and meaning shifts linked by Carey with the fallacies of equivocation and accent.

could be used as a tool to analyze straw man as a fallacy. I set aside even more subtle cases where straw man is combined with related fallacies as a series of projects for future research.[3] We begin with two textbook cases.

2. The Beer and Wine Example

We begin with the beer and wine example, an easy case, and then turn to another textbook example that turns out to be harder but still relatively easy. As noted above, Freeman analyzed the beer and wine example very well by drawing a contrast between the propositions A and B, reprinted below.

A: It would be a good idea to ban advertising beer and wine on radio and television (the concerned citizen's original conclusion).

B: It would be a good idea to get people to stop drinking (the alcohol industry representative's portrayal of that conclusion).

As Freeman wrote, it is far easier to argue against (refute) B than A. It is much easier to cast doubt on B, because it is implausible that B could ever be carried out as a proposal. It simply is not realistic that we could get people to stop drinking by available means, such as persuading them to do so by advertising.

Freeman's analysis of the fallacy in the beer and wine example can be nicely summed up and analyzed using the following argument diagram, constructed with the argument diagramming tool of the Carneades Argumentation System (Gordon, 2005). As shown in Figure 9.1, linked arguments based on argumentation schemes can be displayed in the diagram. Argument from consequences has two schemes, as shown in Chapter 4, Section 3. In argument from positive consequences, a policy or course of action is supported by citing favorable consequences of carrying it out. In argument from negative consequences, a policy or course of action is refuted by citing unfavorable consequences of carrying it out.

In Figure 9.1, the scheme for argument from negative consequences (ANQ) is shown fitting the pro (+) argument on the left. This argument is based on the two premises at the top in the middle supporting the conclusion that it would be a good idea to ban advertising beer and wine on radio and television. But here is also a con argument shown at the bottom, based on the premise that you cannot get people to give up drinking. This premise is also a conclusion of another pro argument shown at the right.

Argument diagramming is shown in this textbook case to be a useful tool for analyzing the structure of the argument in a case where the straw man fallacy has been committed. But it presents only part of the analysis

---

[3] It needs to be added that we cannot exclude such complications entirely and that any adequate analysis of the straw man fallacy has to take account of them.

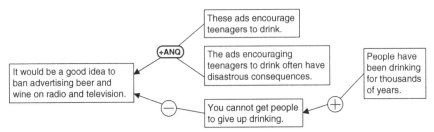

FIGURE 9.1 Argument Diagram for the Beer and Wine Example

needed to prove that the fallacy has been committed. We know from our natural language understanding of the argumentation in the example that the alcohol industry representative wants to make us believe that the concerned citizen advocates total abstinence. But how could we prove this allegation? The answer given by Freeman is evident from how he represents the concerned citizen's position. Freeman (1988, 88) puts the allegation in the form of the question: "Doesn't he suggest that this is implied in that position?" The evidence is that the alcohol industry representative's statement, 'You cannot get people to give up drinking', when placed in the dialogue as his response to the prior move of the concerned citizen, is clearly meant to be a refutation move in the dialogue.

3. The School Prayer Example

Hurley (2003, 122) offered the following example to illustrate the straw man fallacy to readers of his logic textbook. Let's call it the school prayer example. This case is his leading example, used to introduce students to the fallacy by offering a typical instance of it that students can fasten onto to get an intuitive grasp of the error (defined in Section 1, above, by both Hurley and other leading textbook authors).

Mr. Goldberg has argued against prayer in the public schools. Obviously Mr. Goldberg advocates atheism. But atheism is what they used to have in Russia. Atheism leads to the suppression of all religions and the replacement of God by an omnipotent state. Is that what we want for this country? I hardly think so. Clearly Mr. Goldberg's argument is nonsense.

Hurley's explanation (122) is that the fallacy in this case involves two arguments. The first is an argument against prayer in the public schools. The second is an attack on this argument, equating it with an argument for atheism. The conclusion of the second argument is that the first argument is nonsense. Hurley analyzed the fallacy as follows (122). Since the first argument has nothing to do with atheism, the second argument commits the straw man fallacy. This analysis is not entirely accurate, because Hurley's premise is questionable. The first argument does have something to do

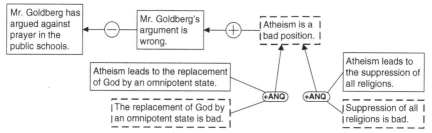

FIGURE 9.2 Argument Diagram 1 of the School Prayer Example

with atheism. Atheists are generally against prayer in public schools, and it is not hard to see why, from what we know of their position. Thus, atheism could certainly be one reason for arguing against prayer in public schools. So what analysis should be given of the school prayer example?

We can diagram the argumentation in this case by first making up a key list, a set of propositions making up the premises and conclusions in the argumentation. First, we have the following four explicit premises.

Mr. Goldberg has argued against prayer in the public schools.

Mr. Goldberg's argument is wrong.

Atheism leads to the suppression of all religions.

Atheism leads to the replacement of God by an omnipotent state.

Atheism is a bad position.

The structure of this first part of the school prayer argument can be visualized using the argument diagram in Figure 9.2. Each implicit premise is represented by a text box with dashed lines around the borders. Two implicit premises have been inserted, as shown in Figure 9.2.

Implicit Premise 1: Suppression of all religions is bad.

Implicit Premise 2: The replacement of God by an omnipotent state is bad.

Each implicit premise functions together with another premise in a pro argument supporting the conclusion that atheism is a bad position. Each argument fits the scheme for argument from negative consequences, as shown in Figure 9.2. In addition, the argumentation scheme for argument from values is involved, since valuation of outcomes as good (positive) or bad (negative) is part of the argumentation, but this part of the argument is not shown in Figure 9.2.

Looking over the argument diagram of Figure 9.2, there is no straw man fallacy committed in this part of the argument. The two linked arguments fit the scheme for argument from negative consequences, leading to the implicit conclusion that atheism is a bad position, which, in turn, when taken as a premise supports the conclusion that Mr. Goldberg's argument is

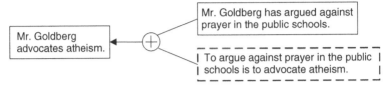

FIGURE 9.3 Argument Diagram 2 of the School Prayer Example

wrong. The premises of this argument are dubious. It is dubious that atheism leads to the suppression of all religions. And one might counter that atheism is a belief about the existence of a deity, not a belief about which actions should be taken toward those with religious beliefs. However, even though the argument pictured in Figure 9.2 is weak and not very plausible, it commits no straw man fallacy. To see where the fallacy occurs, we have to look at a second diagram, in Figure 9.3, showing another part of the argumentation. If we look at the implicit premise on the right at the bottom in this pro argument, we notice the questionable assumption where the straw man fallacy is committed.

Thus Hurley was on the right track when he wrote that two arguments were involved and that because the first argument had nothing to do with atheism, the second argument commits the straw man fallacy. One reason Mr. Goldberg might have for arguing against prayer in the public schools is that he is an atheist or has some atheistic views. But that is not the only reason he might have. He might not be advocating atheism at all. He might be arguing on some other basis, like separation of church and state. Thus arguing against his view by basing the argument on a premise stating that he "obviously" advocates atheism because he has argued against prayer in the pubic schools commits the straw man fallacy. The diagram in Figure 9.3 displays this fault.

To see how the whole argument works, you have to put the two diagrams in Figures 9.2 and 9.3 together. The conclusion of the second one, the proposition that Mr. Goldberg advocates atheism, goes together with the conclusion of the other, the proposition that atheism is a bad position, to lead by another linked argument to the conclusion that Mr. Goldberg's argument is wrong. However, because of the erroneous inference shown in Figure 9.3, the straw man fallacy is committed. Although the argument diagram in Figure 9.3 is very helpful in explaining where the fallacy lies, it does not diagnose the heart of the straw man fallacy. It does not give the analyst hard proof that the argument as displayed commits the straw man fallacy. It tells us only that there is an implicit premise that to argue against prayer in the schools is to advocate atheism. Since we know that this generalization is implausible, we can certainly see, looking at the diagram, why the argument is weak. But the straw man fallacy, according to the definitions in Section 1, is committed in cases where the respondent's position is not the attributed one but a different one (Johnson and Blair's condition 2). However, the argument analyst can prove that this definition fits the argumentation given in the school prayer case, on a basis of burden of proof.

What do we know about Mr. Goldberg's position in this case? We know nothing about it, except that he has argued against prayer in the public schools. We are not told his reasons for so arguing or even whether he offered any. The presumption, then, by default, is that in the school prayer case, we have no information about Mr. Goldberg's position other than what we are told in the case as stated above. On the other hand, it is assumed by the argument attacker that "obviously" Goldberg is an atheist. The use of this indicator word is an indicator of inferability by the addressee (Barbaresi, 1987). The problem is that it is being inferred without proof that Mr. Goldberg's position, the basis for what he advocates, is that of atheism. This attribution of commitment to the position of atheism is, however, unwarranted by the data in the case.

The difference between these textbook cases and a more realistic case like the environmentalist example is that in the latter case, we presumably have quite a bit of information concerning specific statements to which the arguer committed him- or herself in a previous dialogue exchange of views. It is presumed in such a case, and in many realistic cases of the type where a ruling on the straw man fallacy would be useful, that we do know quite a bit about the arguer's position. Thus, it would be possible in such cases, if we had a record of these previous arguments, to go over it and collect evidence on what the arguer committed him- or herself to. Then the argument analyst could compare these actual commitments with the position attributed to him or her by the attacker. This comparison between these two sets of propositions, as shown by Freeman's contrast of the pair of propositions in the beer and wine example, could then function as the evidence to support or refute the allegation that the attacker committed the straw man fallacy. In the school prayer case, we do not have evidence. The assumption on which the allegation of fallacy depends is that there is none – it is assumed that there is no textual evidence of Mr. Goldberg's saying anything that could be an indicator of atheism.

It has been shown in this section how argumentation diagramming is helpful in marking up the argumentation in a given text of discourse to assist in providing an analysis of the argument, revealing some important aspects relevant to deciding whether a straw man fallacy can be said to have been committed. It can represent the premises and conclusion of an argument and, importantly, can represent implicit premises that might be questionable. But examining some other examples will show why this is little more than a beginning.

4. Some Other Kinds of Examples

As always with the study of actual cases, whether in legal argumentation or everyday conversational discourse, there are hard cases as well as easy cases. In a real case such as the Gore and the Internet example, a lot of work had to be done to search out where Gore went on record as making some claim on this issue, and then to check to see what he actually said and what that can be taken to imply about his commitments on the issue. The analysis of the case depends on the assumption that this was all that Gore said. But it also depends centrally on finding what he actually did say, and then comparing these statements with the one attributed to him about having invented the Internet. Proving whether the attackers committed the straw man fallacy can be done only by finding this evidence of what the arguer went on record as saying or writing in the first place. Then it has to be judged which commitments can fairly be attributed to him, based on this text of discourse. Based on these findings, a comparison needs to be made between the commitments and the ones attributed to him by the attacker.

The Gore and the Internet example involves factors such as misquotation, interpretation and taking what was said out of context. What Gore actually said was that he "took the initiative in creating the Internet". What he was later reported as having said, and what he was attacked for, was claiming to have invented the Internet. But it is clear that a commitment to the former proposition does not imply a commitment to the latter, even though it was widely taken to do so. When a careful analysis of what Gore's quoted statement committed him to has been carried out, based on the text of what he actually said, no proposition claiming that he invented the Internet is found. However, as noted in Section 1, what Gore said is ambiguous and vague enough to invite such interpretations by his attackers by making the misattribution seem somewhat plausible, in the absence of the data provided by furnishing the exact details of the quotation. Use of the word 'created' didn't help, because it suggested invention, but taking the initiative in creating something does not imply a claim to creating it. It can be persuasively argued that one can contend by analyzing Gore's commitments in the text that it is indeed an instance of the straw man fallacy. But this example should be classified as a hard case, because to prove the point, the actual wording had to be collected as data, and its implications analyzed.

Among the hardest and most subtle cases are philosophical disputes in which one philosopher is interpreting the writings of another philosopher. Sometimes the one philosopher is merely trying to interpret or clarify the meaning of some doctrine or view attributed to the other. But it is common for such a philosophical exegesis to shade off into criticism or even attempted refutation. It is not possible to attempt to study hard cases of this sort here, even though they are very common. The reason is that philosophical argumentation is typically of such a high order of subtlety that any real case tends to be lengthy and hard-fought. Such cases take us into the areas of metaphilosophy, philosophical criticism and philosophical exegesis.

Judgments about whether or not the straw man fallacy has been committed can be quite subtle, yet quite important to sort out in cases of academic disputes. Particularly in philosophy, much of the work depends on interpreting what some philosopher wrote who is now dead and is no longer available to disavow or correct misinterpretations and refute straw man arguments on the part of those who are now criticizing his or her views. Of course, philosophical argumentation is typically quite abstract, and it is often hard to determine precisely what the philosopher in question was really committed to. If the philosopher wrote in ancient Greek, for example, specialists in Greek language get involved. But even so, when dealing with a dead language, and a different culture that may be remote from our own in important ways, the dangers of attributing views to this philosopher that do not represent what he or she was really trying to say are quite real. In cases like this, what most do is to invoke the so-called principle of charity, a rule of interpretation to the effect that if several versions of the view in question can be read into the text, the most plausible one should be selected. Or if one is trying to select from several candidates, where each could justifiably be said to represent the argument that the philosopher was advocating, then one should select the strongest argument as the one that represents the philosopher's view of the matter.

In his study of Plato's dialectical method, Richard Robinson (1953) gained some insight into the process whereby one philosopher can tend to fall into misinterpreting the views of another philosopher. Robinson (2) warned of five ways in which such misinterpretation is very common. The first way, which he calls "mosaic interpretation", is "the habit of laying any amount of weight on an isolated text or single sentence, without determining whether it is a passing remark or a settled part of your author's thinking, whether it is made for a special purpose or is intended to be generally valid, and so on". The problem in this is to judge how deep a commitment is.

The second type of misinterpretation is called "misinterpretation by abstraction". Robinson (2) describes this form of misinterpretation as follows.

Your author mentions X; and X appears to you to be a case of Y; and on the strength of that you say that your author 'was well aware of Y', or even that he 'explicitly mentions Y'. Because you have abstracted Y from X, you assume that your author did so too. But such an assumption must not be made on general grounds, for no man has ever made or ever will make all the abstractions possible from any one object present to his consciousness.

Context can be very important in judging commitment, and quoting out of context is a tactic often used as a variant of the straw man fallacy.

The third error is called "misinterpretation by inference". In this type of misinterpretation the author says *A*, and *A* implies *B*. It is assumed therefore that the author meant to assert that *B* is true. Robinson (2) pointed out that this conclusion does not necessarily follow, for the author may not have thought that *A* implies *B*. He may not even have been aware of the suggestion that *A* implies *B* (2). The distinction important to assessing cases of this type is that between explicit and implicit commitment.

The fourth kind of misinterpretation is that used to insinuate the future. Robinson described it (3) as the fault of reading into an author's text those doctrines that did not become explicit until later. Perhaps they did not become explicit after the author was long dead. One can see why this kind of misinterpretation is tempting. A commentator may feel that such a process of insinuation of the future is a way of improving what an author said, by making it more "relevant" or making it more up-to-date. The fifth kind of misinterpretation is called (4) "going beyond a thinker's last word". This form of misinterpretation is described by Robinson (4) as ascribing to an author not only all the steps he took in a certain direction but the next step as well.

The pattern suggested by Robinson's classification of the different ways this kind of misinterpretation typically occurs is that the interpreter goes beyond strictly trying to determine what the view of the other party is, based on what he wrote. Instead, he expands the database by bringing in other data, such as what happened later or what the author might have thought if he or she had continued his or her line of thinking further. Thus the conclusions the interpreter draws by inference are not based on the author's real position, based on what he has committed to, based on what he wrote. Such extrapolated attributions are misrepresentations of an arguer's commitments, if they are portrayed as having been drawn accurately from commitments expressed by what he or she wrote or said. Therefore, such cases can rightly be classified under the heading of the straw man fallacy, if the distorted view attributed to the author is used to try to argue against his or her position.

At this point, one might want to invoke something like the principle of charity. This principle rules that one should choose the interpretation that makes the author of the argument appear more "sensible" (Gough and Tindale, 1985, 102). Johnson (2000, 127) formulated this principle as follows: "When interpreting a text, make the best possible sense of it". But how to make sense of a philosophical text may depend on what the author wrote elsewhere, and on what is known about his or her position, even on how he or she changed that position during different stages of writing. His or her viewpoint and attitudes toward subjects related to the text, as far as these are known, could be important. Although the problem of retraction has been studied in Walton and Krabbe (1995), studying hard cases of philosophical argumentation involving retraction will not be attempted here. It is perhaps for this reason that the logic textbooks tend to stay away from such hard cases. The best we can do here is to acknowledge the distinction between hard and easy cases. Such issues of how the straw man fallacy relates to real cases of academic disputes and philosophical argumentation do not appear to have been studied much, in any systematic way. But for all that, they do seem to be quite important. It is impossible to pursue philosophy, or any academic discipline for that matter, without attributing commitments to another party.

The school prayer case makes all this work seem easy by brushing it aside and assuming that nothing else is known about what Goldberg said. Thus the attack on him can be declared a fallacious straw man by default. But are the more interesting real cases like this? It seems likely, judging from the environmentalist and Gore cases, that they are not. Part of the work is to analyze the argument in the case, using an argument diagram or some comparable technique to get a reconstruction of what the premises and conclusions are supposed to be. The other part is to fairly judge what the arguer's commitments really are, as far as can be judged by what he or she said, and what moves he or she made, in previous dialogue exchanges. In some cases, it is hard or even impossible to say what an arguer really is committed to on a given issue. The straw man fallacy is committed, according to the definitions offered in Section 1, only where there is a definite mismatch between what that position was alleged to be, by the other participant in the dialogue, and what it really is. Of course, what his or her position really is can be judged only by his or her record of commitments in the dialogue, based on the given evidence of what the proponent said on the issue before. Thus the concept of an arguer's commitment set in a dialogue should play a key role in determining whether a straw man fallacy has been committed or not in a particular case.

5. Rules for Determination of Commitment in Dialogue

The biggest problem for defining the structure of the straw man fallacy is to build a method for checking to see whether some proposition attributed to an arguer can be found in the position or view that the arguer has expressed in the dialogue between the two parties. The method of analyzing this fundamental notion presented in Walton and Krabbe (1995) is to determine an arguer's position or view using the device of the commitment set of that arguer, derived from Hamblin (1970, 264). Hamblin (1970; 1971) defined a commitment set, also called a commitment store, as a set of statements, for example, a set of statements listed on a blackboard or in a computer database.[4] At each move in a dialogue, statements

---

[4] The term 'statement' is taken as equivalent to the term 'proposition', for the purposes here. A statement is made by means of an assertion and is expressed in a sentence that is true or false.

are inserted into the store or deleted from it, depending on the type of move made and the preconditions and conditions for the type of move, as stated in the dialogue rules. These statements represent what an arguer in a dialogue can be said to be committed to, at any given point in the dialogue, as a result of the past moves or speech acts that that participant has made during the sequence of the dialogue. Thus at any given point in a dialogue, according to this theory, it will be possible to fairly determine what an arguer's set of commitments should be taken to be. The method for determining an arguer's commitments depends on knowing the type of dialogue in which the two arguers are supposedly engaged and the rules appropriate for that type of dialogue at any given stage. Then at a given point in the dialogue, if an arguer makes a move, such as asking a question, making an assertion or putting forward an argument, that arguer's set of commitments are modified according to the commitment rules governing that type of move. They can be modified by the insertion of new statements into the commitment set or by the deletion of statements in a commitment set. For example, if an arguer makes the assertion that snow is white at some point in the dialogue exchange, then on the basis of his or her speech act of assertion, it can be assumed that the arguer is now committed to the truth of the statement that snow is white. Thus the statement 'Snow is white' would be inserted into the arguer's commitment set. If he or she retracts commitment to this statement, the statement is deleted from the arguer's commitment set. The most vexing problem of constructing formal dialogue systems to represent rational argumentation is that of retraction. Much of the effort to analyze commitment in Walton and Krabbe (1995) concentrates on dealing with this problem. The rules for retraction need to vary according to the type of dialogue in which the participants are supposed to be engaged.

It has become evident in the study of fallacies that you cannot get very far by modeling the argument in question as a sequence of reasoning without taking into account the strategic element of how the arguer based his or her argument on what he or she assumed to be the belief or knowledge of the party to whom the argument was directed. To have a rational strategy, the proponent must simulate the beliefs or the position of the respondent, based on what he or she takes to be the respondent's viewpoint or knowledge of the situation. But how can we represent this strategic aspect of argumentation? One way is to use a knowledge and belief model. On this model, the proponent of an argument constructs it on the basis of what he or she knows or believes about what the respondent knows or believes. In other words, simulation is involved. Simulation can be modeled in doxastic and epistemic systems as a form of iterated knowledge or belief. For example, it might be said that agent *a* believes that agent *b* believes that proposition *A* is true. But there are some problems with this approach. One is that such iterated modalities can cease to make intuitive sense once they become very complex. Another is that in order to analyze cases where a fallacy has been alleged to have been committed, trying to figure out what the arguer or his or her listeners actually know or believe can be difficult or impossible. Also, it may not even be necessary.

Another approach is to assemble the textual evidence in the case and use it to make a hypothesis about what the arguer's or listener's commitments are, based on that evidence. Commitment is not necessarily equated with the arguers' actual beliefs, but rather with what propositions they have gone on record as committing themselves to, based on what they said and what is implied by what they said. But how could one judge this, in a given case? The answer is that each given case of an argument needs to be judged in a context of dialogue, or type of rule-governed conversation.

Following the structure of dialogue systems outlined by Hamblin (1970; 1971), a dialogue is seen as a rule-governed normative framework of argumentation. Two parties (in the simplest and most typical kind of case), called the proponent and the respondent, take turns making moves. The rules given for a particular type of dialogue define which kinds of moves are permitted or required and, in particular, which kind of move is permitted or required in response to the last move made by the other participant.

A basic dialogue system called CB (Walton, 1984), one that would now be classified as a persuasion dialogue, can be used to illustrate the basic structure of a Hamblin-style dialogue. There are two parties, the proponent and the respondent. Each has a thesis (proposition) to be proved as his or her ultimate conclusion, and he or she tries to devise strategies to prove this proposition using as premises only propositions that are commitments of the other party. CB is a simple dialogue system that is designed to study how strategies of proof work in persuasion dialogue in cases that involve basic problems of retraction of commitments. There is a nonempty set of rules of inference that can be applied to statements. For its rules of inference CB uses only rules of inference of classical propositional calculus, even though many other defeasible rules of inference of the kind now called argumentation schemes could also be added. *Modus ponens* is a rule, but rules that allow infinite repetitions such as 'S, therefore $S \lor T$' are not allowed. The version of CB used to build commitment query inference engines below will be even further simplified. It will have only one rule, *modus ponens*. In CB a statement T is said to be an immediate consequence of a set of statements $S_0, S_1, \ldots, S_n$ if and only if '$S_0, S_1, \ldots, S_n$, therefore T' is a substitution instance of an inference rule. A statement T is said to be a consequence of a set of statements $S_0, S_1, \ldots, S_n$ if and only if T is derived by a finite number of immediate-consequence steps from immediate consequences of $S_0, S_1, \ldots, S_n$. Here is the complete set of rules for CB (Walton, 1984, 132–135).

Rules of the Dialogue System CB

Locution Rules

(i)   Statements: Statement-letters, S, T, U, ..., are permissible locutions, and truth-functional compounds of statement-letters.

(ii) Withdrawals: 'No commitment S' is the locution or withdrawal (retraction) of a statement.

(iii) Questions: The question 'S?' asks 'Is it the case that S is true?'

(iv) Challenges: The challenge 'Why S?' requests some statement that can serve as a basis in proof for S.

Commitment Rules

(i)   After a player makes a statement, S, it is included in his commitment store.

(ii) After the withdrawal of S, the statement S is deleted from the speaker's commitment store.

(iii) 'Why S?' places S in the hearer's commitment store unless it is already there or unless the hearer immediately retracts his or her commitment to S.

(iv) Every statement that is shown by the speaker to be an immediate consequence of, statements that are commitments of the hearer then becomes a commitment of the hearer's and is included in his or her commitment store.

(v)   No commitment may be withdrawn by the hearer that is shown by the speaker to be an immediate consequence of statements that are previous commitments of the hearer.

Dialogue Rules

(R1) Each speaker takes his turn to move by advancing one locution at each turn. A no-commitment locution, however, may accompany a why-locution as one turn.

(R2) A question 'S?' must be followed by (i) a statement 'S', (ii) a statement 'Not-S' or (iii) 'No commitment S'.

(R3) 'Why S?' must be followed by (i) 'No commitment S' or (ii) some statement T, where S is a consequence of T.

Win-Loss Rules

(i)   Both players agree in advance that the dialogue will terminate after some finite number of moves.

(ii) For every statement S accepted by him or her as a commitment, a player is awarded one point.

(iii) The first player to show that his or her own thesis is an immediate consequence of a set of commitments of the other player wins the dialogue.

(iv) If nobody wins as in (iii) by the agreed termination point, the player with the greatest number of points wins the dialogue, or the dialogue is a draw.

A participant in a Hamblin-style dialogue is not just an entity that makes moves in a dialogue, but is also one to which commitments can be attributed, in virtue of moves made. For example, if a participant asserts a particular proposition, then that proposition is inserted in his or her so-called commitment store. But it could later be removed, if the participant retracts it, by making a move indicating he or she is no longer committed to it. Once a respondent has committed to certain propositions, a proponent can then later use these commitments as premises in valid arguments to get the respondent to become committed to other propositions about which he or she may be doubtful. This process of using rational argumentation based on the respondent's commitment set is the central feature of the formal dialogues that Hamblin constructed for the purpose of analyzing fallacies. Some computer scientists, for example, those working on the design of formal communication languages for multiagent systems, have also adopted the commitment model (Reed, 1998; Singh, 2000).

Statements are inserted into an arguer's commitment set based on speech acts that the arguer has performed. The dialogue rules determine how commitments are inserted based on what the arguer said and what type of speech act it fits. For example, if an arguer asserted proposition P_1, then never retracted P_1 at any later move, he or she may now rightly be held to be committed to P_1. This much seems straightforward. So how could we develop a method or tool that could be applied at any point in a Hamblin-style dialogue to determine whether or not a participant is committed to a specific proposition? It is this kind of tool we need to supplement dialogue models of rational argumentation in order to build an objective method for dealing with problem cases such as those presented by typical cases of the straw man fallacy.

6. Commitment Query Inference Engines

An inference engine is a computer program that is designed to answer a query from a knowledge base. An inference engine contains a database representing facts or statements about some domain, and a set of rules that can be applied to these facts. The inference engine executes the rules, sometimes also referred to as "firing" the rules, by applying them to the facts. Such an inference engine has three main components. The first is an interpreter that carries out the task by applying the rules to the facts. The second is a scheduler that estimates the effect of applying the rules in light of priorities or other criteria. The third is a consistency enforcer that tries to prevent inconsistencies arising as new statements are derived from the facts and rules by the interpreter.

The use of the commitment query inference engine developed below assumes there are two participants engaged in a dialogue of a specific type, such as a persuasion dialogue. Each participant has a commitment set (Hamblin, 1970). This set also includes the central thesis that each party has the goal of proving. Each party also has an inference engine that it can use to make queries about the commitment set of the other party. The problem is to define which kind of engine is best for the job. The simplest engine is the prototype engine E_1. It searches through a commitment store to see if a specific proposition is in the store. Let's call the proposition the engine is searching for '$P*$', meaning it is being questioned whether or not it is a commitment of the participant. The engine searches through each proposition in the store, and if it finds one that matches $P*$, it identifies $P*$ as found. If it searches through the whole set and finds no proposition there that matches $P*$, it answers, 'The other party is not committed to $P*$'. This kind of engine is easy to construct but will not work in all cases. It will find all the explicit commitments of a participant, but it will not find certain kinds of implicit commitments.

A kind of example that poses this problem is the following one. Suppose that the respondent is committed to the proposition P_1 and also to the conditional proposition. If P_1, then $P*$. When the proponent uses E_2 to search through the respondent's commitment store for $P*$, he or she will not find it. However, the respondent would seem to be committed, as least implicitly, to $P*$. The proponent can show that the respondent is commited to $P*$ by applying the deductively valid form of argument *modus ponens* to two statements found in the respondent's commitment store that imply $P*$. A rational arguer is not necessarily committed to all the logical consequences of propositions to which he or she is explicitly committed. But he or she would seem to be at least indirectly committed to ones that follow by one step of inference by a valid rule such as *modus ponens*.[5] How E_2 carries out this task is shown in the sequence of steps illustrated in Figure 9.4.

As Figure 9.4 shows, the commitment query procedure begins by asking the system whether a particular proposition $P*$ is a commitment of the arguer in question. The system then searches through the arguer's commitment store to see if $P*$ can be found. The system gives a yes or no answer to the question, and also furnishes the propositions associated with $P*$. E_2 is useful for dealing with cases where there is a danger of the error Robinson

---

[5] A qualification is necessary. A participant in a dialogue would be committed to propositions following from his commitments by *modus ponens* only if he also accepted *modus ponens* as a rule of inference. However, in CB, both parties accept *modus ponens* as a rule of inference.

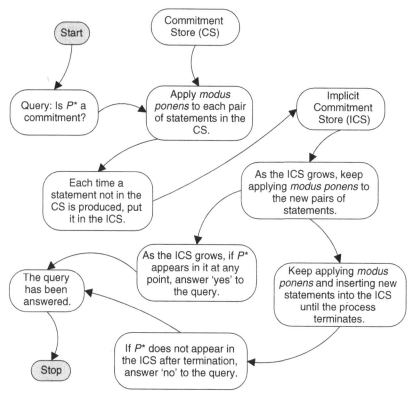

FIGURE 9.4 Sequence of Tasks Carried Out by Inference Engine E_2

called misinterpretation by inference. If a proposition P^* follows logically from some other proposition P_1 that the arguer went on record as committing him- or herself to, then by applying the instructions for implicit commitment search by deductive closure, it can be proved that he or she is committed to P^*. But what if he or she didn't think that P_1 implies P^*? One alternative is to rerun the search engine to check for the proposition 'P_1 implies P^*'. If denial of it is found, that would be quite significant. But if nothing is found, we do not know whether or not he or she is committed to that conditional proposition. The conclusion should be that he or she is implicitly committed to P^*, but not explicitly committed to it. Of course, the respondent can retract such a proposition in CB as a commitment once it has been pointed out to him or her that it follows from some propositions to which the respondent has explicitly committed him- or herself.

We already found in Section 3 that Robinson's third error, called misinterpretation by inference, rests on the important distinction between explicit and implicit commitment. Although E_1 will find all the explicit commitments of any participant in a CB dialogue, it will fail to find some commitments that clearly should be attributed to the agent in a persuasion dialogue. These are implicit commitments that follow from an agent's explicit commitments by forms of argument accepted by both parties as valid. Posing a query to E_2, however, will find the commitments needed to correct this error.

E_2 follows the list of instructions below to answer a commitment query in CB. When you ask E_2 whether the respondent is committed to $P*$, E_2 will start with a pair of propositions it finds in the respondent's commitment set. To answer the query, E_2 will then carry out the following two sets of instructions. X and Y are variables for propositions $P_1, P_2, \ldots, P_n$ that are commitments of an agent.

Instructions for Indirect Commitment Search by E2

Step 1: Apply *modus ponens* to each pair of propositions in the respondent's commitment set to determine whether $P*$ follows from them. Step 1 can be more specifically framed as the following sequence of four steps.

1. $K = \{P_1, \ldots, P_n\}$
2. Form the deductive closure D of K using only *modus ponens*.
3. $K' = K \cup D$
4. If $K' = K$, check to see if $P*$ is in K' else $\{K = K'$; go to step 2$\}$

Step 2: If $P*$ is not yet determined as a commitment, move to the next pair of propositions in the commitment set and apply *modus ponens* to them.
Step 3: Each time a new indirect commitment is found, add it to the commitment store that keeps track of the respondent's commitments.
Step 4: Carry out this procedure recursively, until all the propositions and pairs of propositions in the respondent's explicit and implicit commitment sets are exhausted.
Step 5: If $P*$ is found, answer, "A_1 is implicitly committed to $P*$, because $P*$ follows from commitments X and Y that are either in the respondent's explicit or implicit commitments".
Step 6: If $P*$ is not found, answer, "A_1 is not explicitly or implicitly committed to $P*$ because $P*$ fails to follow from commitments that are either in the respondent's explicit or implicit commitments".

When E_2 follows this list of instructions, it is looking for any indirect commitment that follows from any other proposition, or pair of them, to which A1 is explicitly committed. It does this by applying each of the rules successively to all singletons and pairs of propositions in the respondent's commitment set.

There is a problem with E_2 concerning its failure to find implicit commitments of another sort (Walton and Krabbe, 1995). For example, suppose the respondent has gone on record as arguing, "Socrates is man, therefore Socrates is mortal". He or she is implicitly committed to the proposition, 'If Socrates is a man, he is mortal', because only if we add such an implicit premise is his or her argument valid, and the proposition is a rule that would not likely be in dispute, because it is supported by the common knowledge generalization that all men are mortal. Thus in a persuasion dialogue of the CB type, one party can fulfill its goal of convincing the other to accept the proposition 'Socrates is mortal' only assuming that the proposition 'All men are mortal' is one both should accept as a commitment. This is a classic case of an enthymeme (Hitchcock, 1985), or argument with a missing (implicit) premise.[6] A current problem is to represent arguments that contain implicit premises or conclusions on analyses produced using automated argument diagrams tools (Walton and Reed, 2005). The problem here is that such enthymemes contain premises (or conclusions) that are commitments of an agent, but the search engine E_2 will not identify these propositions as it searches through a commitment set. Is it possible to build a third commitment query engine to deal with cases of enthymemes?

The set of instructions below is designed to construct a set of instructions for a commitment query inference engine that could handle enthymemes. E_3 searches to see if $P*$ can be added to some argument that the respondent has advocated in the previous dialogue that has another proposition as premise to which the respondent is committed. Thus E_3 is searching around to see if $P*$ is a nonexplicit premise in one of the respondent's arguments that the respondent was using to try to prove some conclusion.

Instructions for Implicit Commitment Search by E_3

Step 1: Apply each rule, one at a time, and test to see whether $P*$ is required along with another premise $P1$ of an argument the respondent put forward, where $P1$ is a commitment of the respondent.

Step 2: If $P*$ is not yet determined as a commitment, move to the next proposition $P2$ in the commitment in the set, and apply steps 1 and 2 to it.

Step 3: Each time a new implicit commitment is found, add it to the original set.

Step 4: Carry out this procedure recursively, until all the propositions in the respondent's commitment set are exhausted.

Step 5: If $P*$ is found, answer, "The respondent is committed to $P*$ because $P*$ is an implicit premise or conclusion in an enthymeme containing premise X and/ or conclusion Y".

In this kind of program, E_3 is searching for some argument in a participant's commitment set in which some proposition $P*$ is an implicit component. If it finds such an argument, it will identify $P*$ as an implicit commitment of the agent.

Although E_3 might have its uses, it will not be adequate to search for the kinds of commitments needed to adjudicate on many problem cases such as those arising from problems relating to Gricean conversational maxims (Grice, 1975), the principle of charity and the straw man fallacy. All of these problems involve implicit commitments.

---

[6] We always have to add that it could be a missing conclusion in some instances.

The consistency enforcer is designed to find inconsistent commitments. It searches through a commitment store in the following manner. When it finds any proposition P_1 there, it then searches all through the rest of the store to find the negation of that proposition, not-P_1. If it finds such a pair, it announces, "This commitment set is inconsistent", and it identifies the inconsistency. If it finds no such inconsistency after searching through the whole commitment set, it answers, "This commitment set is consistent". However, on both counts it can be wrong, for reasons similar to this given above in connection with E_3. For an agent might be committed to P_1, but the negation of P_1 might follow from some other proposition, P_3, to which the agent is committed. Thus the problem is similar. E_3 will not find implicit inconsistencies.

E_1, E_2 and E_3 are fairly simple models of inference engines. More sophisticated ones could easily be developed using the many kinds of inference engine technologies available in artificial intelligence (Russell and Norvig, 1995, 92–118). A commitment query system like one of these inference engines is clearly what is needed to deal with the straw man fallacy and related problems of attribution of commitment in argumentation. It will pick up as commitments not only statements to which the agent has explicitly committed itself in the previous dialogue, but also statements so closely related to them that the agent ought be held to be committed to them, too. Now let us go on and see how well it fares in attempting the task of ruling and on real cases where the straw man fallacy has been alleged to have been committed.

7. Looking at the Text and Context of Cases

Cases of argumentation in which the straw man fallacy is thought to have been committed need to be examined on a case-by-case basis. Suppose one arguer has allegedly distorted or misrepresented the position of the other, in order to unfairly attack the other's argument. How can we judge, by verifiable evidence, that the alleged position does not accurately or fairly represent the commitment set of the arguer whose view was attacked? The answer is that we have to look at the text and context of the discourse in the case. Whatever records we have of the arguer's previous speech acts and arguments in the dialogue are the basis of the evidence. Using this given evidence, we can then draw inferences about which statements the arguer can rightly be held to have committed himself to, using the dialogue rules to draw and verify the inferences. This approach works well enough, but which objective system can be used to draw such inferences and to determine the set of commitments that can be taken as the premises?

In the example in Section 1, the argument was, "You want to make this province a heaven on earth, but my objection is that it would leave no room for private property and cause massive unemployment!" The description of this position as wanting to make the province a heaven on earth makes it sound unrealistic and impractical, no doubt an attribution that the representative of the logging industry would need to attack in order to have any credibility with an audience, especially one concerned with costs. But the details of what the representative said in her arguments on the issue are not given. If she never expressed any proposition that could be taken to imply that she wanted to make this province a heaven on earth, the commitment tool could be used to make that finding. This determination would confirm that the argument leveled against her is a straw man fallacy. In the absence of data, however, it cannot be proved.

Let's fill out a few more details of the environmentalist example to get a more specific case. A typical case of the straw man fallacy cited in Walton (1996d, 117) concerns a critical discussion about environmental laws that regulate industrial pollution. One party in the discussion, Bob, has taken what could be called a moderate green position. The other party, Arlene, then argues against Bob's view by saying that he wants to make the earth into the pristine, unspoiled, bucolic place it presumably was before being populated by human beings. Arlene, labeling Bob's view "preservationist", argues that his view is impractical because it is committed to the elimination of private property and industrial manufacturing. Having attributed this extreme preservationist view to Bob, she then goes on to argue that it is impractical, by citing the probable bad consequences of adopting it. She asks the audience to imagine the unemployment and social destruction of private homes that would result if Bob's view were implemented.

The commitment query engine can be applied to this case provided there is enough evidence from the given text of discourse in the case for us to judge what Bob's stated position really is or can be taken to be. In the typical kind of case used to illustrate the straw man fallacy in the logic textbooks, like the school prayer case, little contextual information is presented. The whole argument for the contention that the fallacy has been committed is based on contextual assumptions. In this case, the assumption is that Bob never made any extreme statements that the world should be made into a bucolic place or put forward any other assertions or arguments that could rightly be described as incurring commitment to a preservationist position. Presumably, if all the details of Bob's past arguments and speech acts relevant to the straw man issue were available for inspection and analysis, it could be determined by applying the commitment query system to his commitment set whether it contains any statements committing him to the elimination of private property and industrial manufacturing. If no such statements are there, Arlene committed the straw man fallacy. The same principles of burden of proof identified in the analysis of the school prayer case are operative in this one.

If there was no evidence given, or that can be obtained, that enables application of the commitment search tool to verify the charge leveled against

Bob, then Bob must be assumed to be not guilty. The burden of proof is on Arlene to show, by using the search engine or some comparable objective method of verification, that the commitment attributed to Bob can actually be inferred from his commitment set by a chain of logical inferences. If Arlene cannot offer such evidence to fulfill reasonable requirements of burden of proof, then it can fairly be charged that she has committed the straw man fallacy. Of course, she should have the right to present such evidence and make a case for her allegation. But if the evidence is insufficient, or even nonexistent, then that is evidence she has committed a straw man fallacy against Bob.

As will be indicated in Section 8, it is a limitation of the prototype engines developed in this chapter that they use only propositional logic to determine what follows from commitments. Section 8 discusses how to extend the tools to hard cases where other kinds of reasoning, such as argumentation schemes, need to be considered.

Another type of case often cited in logic textbooks relates to the *ad verecundiam* fallacy. In this kind of case, an expert opinion is cited to back up an argument, but his opinion is misquoted or misrepresented. In the following example, from Wesley Salmon's textbook (1963, 64), an arguer for moral relativism supports his argument by claiming that Einstein proved that everything is relative. Part of the problem with this argument is that it is not justifiable to cite Einstein as an expert on ethical matters, even though he was an expert in physics. The other problem is that is dubious whether Einstein ever really wrote or said anything that can be taken as a commitment to the claim "everything is relative" and is in any sense supporting moral relativism. The main problem with applying the commitment query system to this case is not only that the database would be very large. The problem is to define what counts as finding something in it that counts as saying "everything is relative". Once again, a burden of proof principle needs to be invoked. Unless some specific evidence can be given, on the basis of a commitment search through the database provided by the text of Einstein's writings on physics, the claim that Einstein made such a claim is dubious.

Whatever one might say about this case, it certainly suggests that appeals to expert opinion often run the danger of unfairly distorting an arguer's commitments. Expert opinions cited in support of an argument are often misquoted or otherwise distorted to suit an arguer's needs. In many cases of appeals to expert opinion, the opinion actually made by the expert is not even quoted, as in the Einstein case above. Instead, it is rendered in a simplified form that omits necessary qualifications that limit the scope of the claim (Walton, 1997). The actual wording of what the expert said may be extremely important in judging the commitments implied by it. Once again, this textbook example shows that cases of the alleged committing of such erroneous or deceptive argumentation misdemeanors can be evaluated properly only if the textual data of what was said or written is available for inspection and analysis.

The cases considered so far are not that inherently difficult for the project of applying the commitment query engines to them, even though ruling on them needs to be by default. The problem is not so much one of searching for a specific proposition as one of interpreting what someone said. Thus the commitment search tool has its limits. It does not have the power to disambiguate or clarify something someone said that may be vague or ambiguous. Another problem is that if two interpretations are possible, it does not offer any means of choosing between them or of justifying the choice.

In 2004, Henry Prakken suggested some topics in argumentation theory that are in need of research. As one topic, he suggested studying moves reinterpreting an opponent's position in a dispute in order to better attack it.[7] This remark was highly thought-provoking, as it suggested that such a move could be legitimate in some cases. For example, consider a case where you are confronted with an opponent's argument where different interpretations are possible about what the position really is that the opponent's argument is based on. It is very common to confront an argument that is susceptible to more than one interpretation. Of course, the principle of charity would rule that you should choose the interpretation that makes the argument the strongest. But is that rule always applicable? It can be questioned. After all, the opponent might have meant to put forward the weaker interpretation. Although it may be harder for him or her to defend, it might be the interpretation that he or she really meant or that best does justice to his or her position as a coherent whole. Here, picking the argument that is the easier one to refute might not be so bad.

Or consider the kind of case where you and your opponent are engaged in a persuasion dialogue in which there is a conflict of opinion on some central issue. Each side has the goal of proving its thesis by using the strongest possible rational arguments and of refuting or attacking the opponent's arguments by using the strongest possible criticisms and counterarguments. Now suppose in a context of this sort your opponent comes up with an argument that is open to two different quite legitimate interpretations, one that makes his or her argument stronger and the other that makes it weaker. Each argument is based on a different position. One position is such that you can more easily refute it than the other. What should you do here?

Of course, the obvious thing to do is to ask the opponent which position he or she means to take or to apply the commitment search tool to check for instances where a commitment to the one view or the other has been expressed or implied. If he or she is clearly committed to one, you can

---

[7] Some of Prakken's views on these matters are discussed in Walton (2003).

attack that one. But suppose the opponent does not reply or for whatever reason refuses to choose the one interpretation over the other, and the search tool gives no indication of commitment to the one or the other. What should you do next? Well, in a critical discussion, you are supposed to try to win by using your strongest arguments and by attacking your opponent's arguments at their weakest points, trying to undermine them if possible. That's part of the central communal goal of the critical discussion as a type of dialogue. Such a dialogue can be successful only if both parties bring forward the strongest argument they can and attack the opponent's arguments as forcefully as they can. Its goal is to reveal the strengths and weaknesses of the positions on both sides as fully as possible so that light can be thrown on the issue.

For these reasons I began to wonder if the straw man fallacy is as straightforward as it has seemed in the past. It seems that reinterpreting an opponent's position in order to better attack it might be a fairly common argumentation practice, and not an unreasonable one at that. Consider the kind of argumentation that takes place in a trial, for example. Your job as an advocate is to present the strongest case supporting your client's plea, and part of that is to attack the opposing counsel's argument by finding the weak and questionable points in it. You should even try to refute it, if possible. How should you proceed if the opponent's argument can be interpreted in different ways, one making it stronger and one making it weaker and easier to attack? Clearly, you should choose the latter interpretation. If the opposing counsel disagrees with that interpretation, it is up to him or her to make clear that it is not based on the position he takes.

8. Additional Components of the Commitment Search Tool

The commitment query inference engines E_1, E_2 and E_3 are simple in their design. By themselves, they cannot present a complete analysis that proves mechanically that the straw man fallacy has been committed or that it has not, at least in the hard cases. In this last section, six additional components that need to be added to this tool to make it adequate for this purpose are described. Each of these components has already been extensively studied in argumentation, computer science or linguistics. In some instances, computer programs, artificial intelligence techniques or argumentation methods provide resources for the component that has already been developed.

As shown in the definition of the straw man fallacy in Section 1, for the fallacy to be committed, it has to be assumed that there are two parties taking part in a dialogue and that one is attempting to refute the argument of the other. This requirement is not trivial. It involves the notion of rebuttal, or attempted refutation. How could one prove that such a rebuttal is present in a given text of discourse representing a case? This poses the problem of how to identify an argument, as well as a secondary argument that is supposed to be rebuttal of a given argument in a text of discourse. Of course, we are all very good with our intuitive skills of recognizing an argument when we see one. And there are argument indicators used in informal logic that can aid in such identification tasks. Indeed, the technique of argument diagramming applied to the easy cases in Section 2 aided in this task by helping an analyst to mark up premises, conclusions and refutations. But we do not yet have automated tools to carry out such analysis without a human user who understands natural language. Nor do we yet have automated tools that would enable a computer search system to scan through data, such as documents on the Internet, and identify specific arguments or even types of arguments. So here is one additional component. It could be called the refutation component.

In some cases, the database may be huge. The commitment query inference engine may have to be applied to a body of writings, like a series of books and published chapters or a lengthy trial transcript. The hard cases studied above presented a wide range of problems, but surely one of them is that of doing the work required to collect all the data needed to arrive at a fair judgment of the evidence. In the cases of interpretation and criticism of philosophical arguments we considered, a fair critic may have to examine what the philosopher wrote by collecting and comparing different passages, looking at the actual wording with care. In the Gore and Internet example, the main problem is that none of the critics who initially made the allegation did his or her homework. What is needed to deal with this task is a database component for collecting sufficient information about a case.

A third, additional component needed is a device for dealing with examples where the given text of discourse is incomplete. One of the simplest problems was that in the short examples typical of the kinds of cases found in logic textbooks, there is not enough textual data given in the description of the case to completely prove the allegation that the arguer has committed the straw man fallacy or has not. The example in such a case is only a short quotation from an article or even a book. There may simply be not enough of a database to which to apply the search engine. The component needed here is a device to draw a conclusion from an incomplete database so that a provisional judgment can be made, subject to reappraisal if more details of the case are collected, analyzed and taken into account. Such a device could be called the closure component.

A limitation of the commitment query inference engine models set out above is the restriction to deductive logic. The tool could potentially be much more useful if the inference engine could be expanded so that it contained not only deductive rules of inference, but also inductive ones, and even defeasible argumentation schemes. The addition of this capability could provide a much more sophisticated search engine for implicit commitment determination. Such a tool can best be seen as being part of a system of dialogue in which the argumentation of each party is based on the commitments of the other. Thus the tool is best seen as a dialectical device for prompting questions such as "Are you really committed to $P*$?" or "Do you realize that your explicit commitment to P_1 provides evidence that you are committed to $P*$?"

The three inference engines modeled above use only propositional logic to define what follows from an explicit commitment or set of them by logical inference. They can be extended to define logical consequence for other deductive types of argument, such as first-order logic with quantifiers. They can verify the type of fault Robinson called "misinterpretation by inference", where the author says A and A deductively implies B, by such deductive criteria. There is still the problem cited by Robinson (1953, 2) of whether the author really meant to assert that B is true, for he or she may not have thought that A implies B. Still, in many instances, depending on the context of dialogue, if the author said A, and A implies B by propositional logic, or some extension in deductive logic, it might fairly be alleged that he or she can be held to B as an implicit commitment in the dialogue. Here the distinction between an explicit and an implicit, or derived, commitment is important to recall. When applied in a framework of dialogue such as CB, engine E_1 can always decisively answer the question of whether or not a participant is committed to a proposition $P*$. But in cases of implicit commitments, either E_2 or E_3 needs to be applied. But even there, it needs to be recognized that these two kinds of commitments have to be handled differently. Once an arguer's implicit commitment is pointed out to him or her, by deriving it from a pair of his or her explicit commitments, the arguer may decide to retract all three as commitments, saying that now he or she sees the implicit commitment for what it is, that he or she should not be committed to it, because it does not really represent his or her position in the dialogue as a whole.

In this light, we now return to consider Robinson's method of mosaic interpretation, which he defines as the habit of laying any amount of weight on an isolated text or single sentence, without determining whether it is a passing remark or a settled part of the respondent author's thinking. The commitment query search engine may find a proposition indicating the commitment sought, but the problem here is to determine how deep a commitment it is. Is it just a chance remark, or is it meant by the author to represent an important new insight that is essential to his or her viewpoint? This additional query suggests that applying the commitment query search engine should be seen as just part of a broader dialectical procedure for the determination and judging of commitment. In addition to a simple commitment check of the kind carried out by search engine E_3, there may need to be some analysis of the place of the proposition found as part of an arguer's broader and deeper commitment to a view. If the author can be questioned, that should be the method to use. But even if not, comments he or she made may be evidence of this proposition in his or her position. Or there may be some clash between the statement found by applying the query engine to his or her commitment set and his or her thesis to be proved (ultimate conclusion) in the dialogue. In such a case, the use of any one of the commitment query search engines needs to be supplemented by also taking into account other factors in the dialogue. The use of the search engines needs to be embedded in a broader dialogue framework such as CB.

Even where a database is large enough and has been carefully analyzed by the above tools, there can remain problems of trying to interpret what some potentially ambiguous or vague utterance of an arguer means. In the Gore and the Internet example, part of the problem was that Gore suggested that he claimed to have created the Internet, even though collecting the real text of what he said later showed the suggestion to be misleading and inaccurate. Such cases need to be decided on a burden of proof basis by studying the context of dialogue and applying Gricean conversational principles and normative rules for interpretation, such as the principle of charity. Both an implicature component and a dialogue component are required.

Philosophical argumentation can contain very hard cases. Philosophers often contest conflicting interpretations of the viewpoint of another philosopher long dead. It is common to encounter philosophical books of this sort. Very fine points of interpretation may be contested, based on the textual evidence and on what it may be taken arguably to imply. The author may have changed his or her mind several times and made retractions. Or what he or she said in one place may be inconsistent with what he or she said in another place, and the author may never have resolved or even apparently noticed the inconsistency. In such hard cases, the search engine by itself is insufficient, because the context of dialogue needs to be taken into account. Legal arguments are also often among the very hard cases. Decisions arrived at by courts are often based on contested statutory interpretations. A legal case at trial may have to be treated as a special kind of case because an advocate's job is to make a persuasive case and present as strong an argument as possible for one viewpoint. The principle of charity may not apply. Thus the commitment search tool needs to be seen as just one part of a wider dialectical method that may have to deal with competing interpretations of a text that is vague or ambiguous. Still, the method itself is useful, even though applying it to real cases may require amassing a lot of data from the text of discourse in a case and use of conversational rules to interpret the data. What is needed here is a dialectical component. In analyzing a straw man argument in an example, some key assumptions about the type of dialogue may be required. In particular, the question of which dialogue rules apply may affect the evaluation of the argument.

Which dialogue rules should govern the incurring and retraction of commitments? The rules for permissive persuasion dialogue (PPD) in Walton and Krabbe (1995, 133–140) define a dialogue system for managing commitments in persuasion dialogue that is much more complex than CB. There is an initial conflict description giving the initial set of commitments of both the proponent and respondent (133). The commitment rules determine the commitments of each party after each move (134). Each move is a six-tuple containing any or all of the following components: retractions, concessions, requests for retractions, requests for concessions, arguments and challenges (135). These components can be called speech acts. There are, in sum, eighteen rules governing how each party must react when a move of a certain type was made. But PPD has many rather complex features, such as the use of dark-side commitments that involve matters beyond the scope of the present discussion. To study problems of commitment management in relation to the examples discussed above, we began with a simpler model of dialogue than PPD, namely, CB. In CB, as in PPD, there is an initial conflict description and a set of commitment rules. But there are no dark-side commitments. And each party is allowed to put forward only one type of component (speech act) at each move. For example, a participant can put forward an argument, and then the other participant has to respond to that argument by challenging it, asking a critical question or making whatever is considered an appropriate response.

When it comes to modeling argumentation of the kind associated with informal fallacies, consideration of one type of dialogue is not adequate for all cases. A rigorous persuasion dialogue (RPD) was another model studied in Walton and Krabbe (1995). In this type of dialogue, each participant can put forward only one speech act at each move. The rules determine that at the next move, the other party must choose from a limited range of options. In addition, there are other types of dialogue to be considered, such as the inquiry and information-seeking types. The rules for these types of dialogue are different from the ones appropriate for a persuasion dialogue. Questions can be raised on how rigorous or permissive the rules for persuasion dialogue should be.

When applying the inference engine models to the examples of the straw man fallacy studied in the latter sections of this chapter, six additional components were brought into consideration: (1) the refutation component, (2) the database component, (3) the closure component, (4) the implicit commitment component, (5) the implicature component and (6) the dialogue component. It is beyond the scope of this chapter to attempt to develop any of these components any further. What has been demonstrated is that the central tool for analysis and evaluation of the straw man fallacy is the commitment query inference engine. The commitment query engine could potentially be used for many purposes. It could be applied to analyzing and evaluating arguments even where no allegation of committing the straw man fallacy has been made or is appropriate. Still, examples where such an allegation has been made provide excellent test cases for instructing us on how to develop commitment query search engines as useful argumentation tools.

9. Are There Variants of the Straw Man Fallacy?

Aiken and Casey (2011) distinguished three forms of the straw man fallacy called straw man, weak man and hollow man. Moreover, what they call the straw man fallacy is said to have two forms. In what they called the simple version of the fallacy (89), the proponent caricatures or distorts the respondent's position and thereby attributes a significantly less defensible position to him or her. Then the proponent criticizes this less defensible position, argues that this position is wrong and concludes that the respondent is wrong.

Within the first category, called straw man, there is another version of the straw man fallacy said to be subtly different from the simple one. In this "more subtle" version (89), the proponent, instead of attacking the respondent's position directly, attacks some arguments that the respondent has used to support his or her position but misrepresents these arguments. The distinction drawn in this regard is between an arguer's position and one or more arguments used by the arguer to support that position. It would appear that Aiken and Casey understand the notion of an arguer's position as referring to the arguer's commitment set in the sense of Hamblin (1970), as recommended in the preceding sections of this chapter, although I cannot find any statement in their paper where they make this claim, nor is Hamblin's book found in their list of references.

The second form of the straw man fallacy postulated by Aiken and Casey (89), called weak man, is committed where the respondent has put forward several arguments to support his or her position, and the proponent selects the weakest of these arguments, refutes that argument and claims to have defeated the respondent's overall case for his or her position. The third form of the straw man fallacy, called hollow man (92), is committed where the view taken to represent the respondent's position is not merely a caricature of that position but a complete fabrication. In the case of the hollow man fallacy, the position actually attacked represents no particular discussion and bears no relation to any view expressed in the discussion: "the hollow man consists in fabricating an imaginary opponent with an impossibly weak argument, and then defeating the argument" (92).

What Aiken and Casey call the weak man fallacy is problematic, precisely because of the dialectical variability of the straw man argument. Aiken and Casey (88) made the important point that straw man is a dialectical fallacy. As they put it, straw man is not really an invalid form of argument that fails to support a claim made. They describe it as a broader failure that occurs in a dialectical context where one arguer challenges another but distorts some significant feature of the opponent's case. This fundamentally important point that has been the basis of the analysis of the straw man fallacy in this chapter is a dialectical failure to use the commitment set of an opponent properly in a dialogue by setting up a straw man that does not accurately represent that opponent's position. However, this point also opens the possibility of dialectical variability, in turn opening the possibility that such a move could be legitimate in some cases, depending on the context of dialogue in which the argument is used.

For example, if a professor in an introductory class is representing a very subtle philosophical position to the class, it may be permissible for him or her to start the discussion by presenting an oversimplified account of that position that, at least for the present, overlooks qualifications to it that need to be eventually brought out in the discussion. When the professor puts forward counterarguments to this position that the class will discuss, it may be argued that he or she is committing a form of the straw man fallacy (Ribeiro, 2008). It can be argued that making the position simpler, even if only for purposes of discussion, is distorting it and that the arguments against that position that are being considered commit the straw man fallacy. The example is a tricky one, because the dialectical framework is complex. It may be that the professor is not merely trying to refute the position he or she is discussing but is rather trying to explain it to the class by getting the students to appreciate the arguments for and against it that can be brought forward.

There is much more to be said here about dialectical variability that is especially relevant to the so-called weak man variant of the straw man fallacy. It can be questioned whether it might be permissible, or even recommended in some cases, for an arguer to ignore his or her opponent's strongest argument and concentrate on attacking one of the opponent's weaker arguments for his or her position. It is easy to appreciate this point if you look at legal argumentation in a trial, let's say in a criminal case, where the prosecution has put forward its argument and the defense needs to devise a strategy to reply. Should the defense attack the prosecution's strongest argument or one of its weaker arguments? This decision will depend on a number of factors. In many cases it may well be the best strategy for the defense to leave the prosecution's strongest argument alone and to concentrate on attacking one of his or her weaker arguments that is more vulnerable.

The reason is that in a critical discussion, when you attack an opponent's argument, there has to be room for strategic maneuvering so that you are allowed to select one of the arguments he or she has used to defend his or her position and to attack this defense. It needs to be up to the proponent to make a decision on which of these arguments to select, because it is an important feature of the critical discussion, or persuasion type of dialogue, that each party be allowed the freedom to probe into the position and the supporting arguments of the other side in order to find the weaknesses in them that most need to be critically scrutinized (van Eemeren, 2010). This selection depends on the proponent's strategy, and he or she may typically need to devise a plan to probe into the opponent's position and supporting arguments at a weak point that may not be obvious but that can be exploited by putting forward a series of critical questions, requests for clarification and counterarguments. In short, just because the proponent looks over several arguments put forward by the respondent to defend his or her position, and selects the weakest of these as the focus of his or her critical questioning and counterarguments, it by no means follows that the proponent has committed a straw man fallacy. It may just be a good strategy.

Lewinski (2011, 491) has shown that there is a dialectical variability involved in evaluating cases where a straw man fallacy has allegedly been committed. This aspect of dialectical variability means that the argument evaluation needs to take into account context-specific rules of interpretation and commitment attribution. He provides evidence for this thesis by contrasting two kinds of cases. For example, one type of case could be a legal trial or a peer academic review where it is necessary for an arguer to stick strictly to the literal meaning of the expressions used by the party who is criticizing. Any deviation from sticking to the explicit, overt meaning of what was said is limited by the requirements of precision. In the legal cross-examination, for example, the party being examined needs to expect strong criticisms designed to expose weaknesses in his or her argument. Now contrast cases of this sort with ordinary discussions of a looser and conventionally polite sort where there is a preference for agreement and a general cooperative spirit. Consider, for example, a chat in a pub about football or the current political situation. The expectations and requirements for precision of quoting a previous remark and sticking to its precise literal meaning can be very different in the two dialogical settings. Dialectical variability can be taken into account by the variability in the formal PPD and RPD models of dialogue explained in Section 8 above.

Once dialectical variability is taken into account, it becomes much harder to prove that an argument in a particular case is an instance of the weak man fallacy. The reason is that the rationale behind classifying a weak man argument as a fallacy is much less powerful than the rationale behind classifying the basic form of straw man argument (what Aikin and Casey call the simple form of the straw man fallacy) as a fallacy. The importance of this point can be appreciated by asking why it is fundamentally important for each party in a persuasion dialogue to be allowed the freedom to probe into the position of the other side. The reason is that persuasion dialogue is a type of dialogue that proceeds by each party critically examining the position of the other party and using the commitments of the other party as premises in building its arguments. Persuasion dialogue is an adversarial type of dialogue in which each party has the goal of presenting arguments to prove its own thesis, where each side has a thesis to be proved, where this burden of proof (or burden of persuasion, as it is called in law) is set at the opening stage. The term 'persuasion' is used in a technical and normative sense, referring to the goal of rational persuasion that each party has in this type of dialogue. The proponent has the goal of proving his or her thesis by means of a chain of arguments composed exclusively of premises to which the respondent is committed. The respondent has the goal of proving his or her thesis by means of a chain of arguments composed exclusively of premises to which the proponent is committed. A successful act of persuasion consists of proving your own thesis from a set of premises that are all commitments of the other party. This way of configuring the structure of a persuasion dialogue shows why it is a centrally important rule of this type of dialogue that a rational arguer should never misrepresent or distort the commitments of the other party in arguments he or she constructs to prove his or her thesis. The reason is that committing this kind of error goes directly against the central purpose of dialogue, which is one of rational persuasion.

In short, there is a very strong rationale for the simple version of the straw man fallacy, making it a very important failure in the persuasion type of dialogue, while in the case of the weak man fallacy, there is no comparable strong rationale in persuasion dialogue. While it may represent an interesting kind of argument that can go wrong or be exploited by attempting to unfairly get the best of the speech partner, it is much harder to diagnose and prove to be fallacious as a move in a dialogue, compared with what Aikin and Casey call the simple form of the straw man fallacy. The simple form of the straw man fallacy is more centrally important, because it is a rule of critical discussion that when the proponent attacks the position of the respondent, he or she must accurately represent that position by drawing on the record of statements actually made by the respondent and recorded in the respondent's commitment set. If the proponent commits the straw man fallacy by distorting the respondent's position so that it represents commitments that the respondent has not actually made in the dialogue, as shown by the evidence of the previous moves he or she has made in the dialogue and the propositions he or she has accepted, this failure is definitely a serious problem and rightly classified as an important informal fallacy. The reason is that it undermines the central purpose of a persuasion dialogue, which is to resolve the conflict between two positions by having the advocate of each of the positions support his or her position based on premises that are commitments of the other side. The ultimate aim of the dialogue as a whole is to determine which of the two positions has best survived the critical scrutiny to which it has been subjected during the discussion, so it can be determined which side has won.

The hollow man fallacy is not a problem as long as we define it more narrowly as a version of the simple straw man fallacy that is an extreme variant in which the position attacked by the proponent is a mere caricature that is so far removed from his or her real position that it bears no relation to it at all. Seen in this way, the hollow man fallacy is a subspecies of the simple type of straw man fallacy, and it has the same kind of rationale as the simple type of straw man fallacy in a persuasion dialogue. The only problem here is that of drawing a clear line between cases of the simple straw man fallacy and the hollow man variant. However, the hollow man fallacy becomes a serious problem to analyze if we construe it, as Aikin and Casey construe the weak man fallacy, as a fallacy that involves not just distorting the opponent's position, but also choosing his weaker arguments for that position to attack.

10. Other Fallacies Related to Straw Man

There remains the problem of distinguishing the straw man fallacy from other kinds of fallacies and misdemeanors of argumentation to which it is closely related. One of these is the fallacy of *secundum quid*, or neglect of qualifications, treated in Chapter 8, Section 7. One way to misrepresent an opponent's position in political argumentation, for example, is to make it appear to be more extreme by omitting necessary qualifications. This more extreme version of the position is much easier to refute. Straw man is also closely related to the use of misquotations in argumentation and to the fallacy of wrenching from context.

Misquotation has been a substantial problem in many legal trials. To cite one example from Walton and Macagno (2011, 40), in the trial of Galileo, some quotations used against him were manipulated by changing words with the result that his position was unfairly represented as evidence against him. Two words had been changed, and this misquotation made Galileo appear to be guilty of the accusations made against him. This case and the others cited by Walton and Macagno (2011) appear to fall into the straw man category. The reason is that the incorrect changing of the wording had made Galileo's position appear to be more open to the counterarguments put forward against him in his trial. Yet it is clearly and demonstrably a case of misquotation.

In cases such as this we appear to be on the borderline between the fallacy of straw man and a failure of misquotation, which may appear to be a different kind of fallacy or improper argumentation move. However, this problem is easily solved by seeing that misquotation is just one of the many means of committing the straw man fallacy. Another one of these means is that of wrenching from context. But the problem is more acute here, because wrenching from context is widely recognized in logic textbooks and other writings on argumentation as a distinct fallacy in its own right.

Twenty examples presented in Walton and Macagno (2010, 285) have been used as evidence to support the conclusion that wrenching from context is a distortion of the other party's position in the discussion arising from the emphasis on a particular aspect of a quotation. However, it is also commented (285) that there are blurred boundaries between wrenching from context and the straw man fallacy that make it a less than straightforward task to classify these different kinds of manipulative moves as species of fallacy or argument failure.

It is concluded here generally that the fallacy of wrenching from context is best seen as a fallacy of commitment. That has been the central claim of this chapter. However, it is also concluded that the straw man fallacy is different from misquotation and wrenching from context *simpliciter* because it is based on a misrepresentation that is always used to attack the commitment of the other party by distorting it in order to make it easier to refute. Wrenching from context is based on a contextual shift that needs to be explained by employing Gricean implicature (Walton and Macagno, 2010). However, the qualification needs to be emphasized that the straw man fallacy can sometimes be based at least partly on misquotation, and it can also in some cases be based on wrenching from context. It follows that there are cases of arguments in which both fallacies (straw man and wrenching form context) are committed and also where misquotation is part of the problem. Nevertheless, the two fallacies are conceptually distinct and are distinct from misquotation as a phenomenon of argumentation.

For the present, the most central problem should be that of finding the requirements for correctly determining an arguer's position by the evidence that can be collected from a set of commitments in a dialogue and any other evidence that is available. This problem was shown in this chapter to be a technical problem solvable by computer science – the problem of building an inference engine that could search through the evidence in order to determine fairly whether or not a given commitment can properly be said to be part of an arguer's position. Once this technical problem is solved, it will be a giant step forward for analyzing the straw man fallacy, and it will provide a firm basis for moving ahead with further argumentation studies to examine borderline cases so straw man can be clearly distinguished from other kinds of fallacies and problematic argumentation moves. This problem has already been solved for search engines using deductive reasoning, but to make it applicable to the straw man fallacy, it needs to fit into a system that can use defeasible argumentation schemes, for example, the Carneades Argumentation System. This is an important project for future work in computational argumentation studies.

Bibliography

Aikin, S., and Casey, J. (2011). Straw Men, Weak Men and Hollow Men. *Argumentation*, 25(1), 87–105.

Aleven, V. (1997). *Teaching Case Based Argumentation through an Example and Models.* Ph.D. thesis, University of Pittsburgh.

Aristotle (1928). *On Sophistical Refutations.* Trans. E. S. Forster, Loeb Classical Library, Cambridge, Mass.: Harvard University Press.

Aristotle (1937). *The Art of Rhetoric.* Trans. John Henry Freese, Loeb Classical Library, Cambridge, Mass.: Harvard University Press.

Aristotle (1939). *Topics.* Trans. E. S. Forster, Loeb Classical Library, Cambridge, Mass.: Harvard University Press.

Ashley, K. (1988). Arguing by Analogy in Law: A Case-Based Model. In *Analogical Reasoning*, ed. D. H. Helman. Dordrecht: Kluwer, 205–224.

Ashley, K. (2004). Capturing the Dialectic between Principles and Cases. *Jurimetrics*, 44, 229–279.

Ashley, K. (2006). Case-Based Reasoning. In *Information Technology and Lawyers*, ed. A. R. Lodder and A. Oskamp. Berlin: Springer, 23–60.

Ashley, K. (2009). Ontological Requirements for Analogical, Teleological and Hypothetical Reasoning. In *Proceeding of ICAIL 2009: 12th International Conference on Artificial Intelligence and Law.* New York: Association for Computing Machinery, 1–10.

Atkinson, K., Bench-Capon, T.J.M., and McBurney, P. (2004). Justifying Practical Reasoning. In ed. F. Grasso, C, Reed and G. Carenini, *Proceedings of the Fourth International Workshop on Computational Models of Natural Argument* (CMNA 2004), Valencia, Spain, 87–90.

Atkinson, K., Bench-Capon, T.J.M., and McBurney, P. (2005). Arguing about Cases as Practical Reasoning. In *Proceedings of the 10th International Conference on Artificial Intelligence and Law*, ed. G. Sartor. New York: ACM Press, 35–44.

Ballnat, S., and Gordon, T. F. (2010). Goal Selection in Argumentation Processes. In *Computational Models of Argument: Proceedings of COMMA 2010*, ed. P. Baroni, F. Cerutti, M. Giacomin and G. R. Simari. Amsterdam: IOS Press, 51–62.

Barbaresi, L. M. (1987). Obviously and Certainly: Two Different Functions in Argumentative Discourse. *Folia Linguistica*, 21, 3–24.

Bench-Capon, T.J.M. (2003). Persuasion in Practical Argument Using Value-based Argumentation Frameworks. *Journal of Logic and Computation*, 13, 429–448.

Bench-Capon, T.J.M. (2009). Dimension Based Representation of Popov v. Hayashi. In *Modelling Legal Cases*, ed. K. Atkinson. Barcelona: Huygens Editorial, 41–52.

Bench-Capon, T.J.M. (2012). Representing Popov v. Hayashi with Dimensions and Factors. *Artificial Intelligence and Law*, 20, 67–76.

Bench-Capon, T.J.M., and Dunne, P. E. (2007). Argumentation in Artificial Intelligence. *Artificial Intelligence*, 171, 619–641.

Bench-Capon, T.J.M., and Prakken, H. (2010). Using Argument Schemes for Hypothetical Reasoning in Law. *Artificial Intelligence and Law*, 18(2), 153–174.

Bench-Capon, T.J.M., Lowes, D., and McEnery, A. M. (1991). Argument-Based Explanation Logic Programs. *Knowledge-Based Systems*, 4(3), 177–183.

Bex, F. (2009a). Analysing Stories Using Schemes. In *Legal Evidence and Proof: Statistics, Stories, Logic*, ed. H. Kaptein, H. Prakken and B. Verheij. Farnham: Ashgate, 93–116.

Bex, F. (2009b). *Evidence for a Good Story: A Hybrid Theory of Arguments, Stories and Criminal Evidence*. Ph.D. thesis, University of Groningen.

Bex, F. (2011). *Arguments, Stories and Criminal Evidence: A Formal Hybrid Theory*. Dordrecht: Springer.

Bex, F., and Prakken, H. (2010). Investigating Stories in a Formal Dialogue Game. In *Computational Models of Argument: Proceedings of COMMA 2008*, ed. P. Besnard, S. Doutre and A. Hunter. Amsterdam: IOS Press, 73–84.

Bex, F., and Walton, D. (2010). Burdens and Standards of Proof for Inference to the Best Explanation. In *Legal Knowledge and Information Systems: Proceedings of JURIX 2010*, Amsterdam: IOS Press, 37–46.

Bex, F., Prakken, H., Reed, C., and Walton, D. (2003). Towards a Formal Account of Reasoning about Evidence: Argumentation Schemes and Generalizations. *Artificial Intelligence and Law*, 11, 125–165.

Bex, F., Bench-Capon, T.J.M., and Atkinson, K. (2009). 'Did He Jump or Was He Pushed?': Abductive Practical Reasoning. *Artificial Intelligence and Law*, 17, 79–99.

Black, E., and Hunter, A. (2007). A Generative Inquiry Dialogue System. In *Sixth International Joint Conference on Autonomous Agents and Multi-agent Systems*, ed. M. Huhns and O. Shehory, 1010–1017.

Black, E., and Hunter, A. (2008). Using Enthymemes in an Inquiry Dialogue System. In *7th International Joint Conference on Autonomous Agents and Multiagent Systems (AAMAS 2008)*, ed. L. Padgham et al. Estoril, Portugal, May 12–16. Vol. 1, pp. 437–444.

Blair, J. A., and Johnson, R. H. (1987). Argumentation as Dialectical. *Argumentation*, 1, 41–56.

Bondarenko, A., Dung, P. M., Kowalski, R. A., and Toni, F. (1997). An Abstract Argumentation-Theoretic Approach to Default Reasoning. *Artificial Intelligence*, 93, 63–101.

Bratman, M. (1987). *Intention, Plans and Practical Reason*. Cambridge, Mass.: Harvard University Press.

Brewer, S. (1996). Exemplary Reasoning: Semantics, Pragmatics and the Rational Force of Legal Argument by Analogy. *Harvard Law Review*, 925, 923–1038.

Burke, M. (1985). Unstated Premises. *Informal Logic*, 7, 107–118.

Burnyeat, M. F. (1994). Enthymeme: Aristotle on the Logic of Persuasion. In *Aristotle's Rhetoric: Philosophical Essays*, ed. D. J. Furley and A. Nehemas. Princeton, N.J.: Princeton University Press, 3–55.

Caminada, M.W.A. (2008). A Formal Account of Socratic-style Argumentation. *Journal of Applied Logic*, 6(1), 109–132.

Carey, S. (2000). *The Uses and Abuses of Argument*. Mountain View: California, Mayfield.

Clark, K. L. (1978). Negation as Failure. In *Logic and Data Bases*, ed. H. Gallaire and J. Minker. New York: Plenum Press, 293–322.

Cooke, E. (2006). *Peirce's Pragmatic Theory of Inquiry: Fallibilism and Indeterminacy*. London: Continuum.

Copi, I. M. (1986). *Introduction to Logic*, 7th ed. New York: Macmillan.

Copi, I.M., and Cohen, C. (1994). *Introduction to Logic*, 9th ed. New York: Macmillan.

Copi, I. M., and Cohen, C. (1998). *Introduction to Logic*, 10th ed. Upper Saddle River, N.J.: Prentice Hall.

Costantini, S., and Lazarone, A.(1995). Explanation-Based Interpretation of Open-Textured Concepts in Logical Models of Legislation. *Artificial Intelligence and Law*, 3, 191–208.

Dung, P. M. (1995). On the Acceptability of Arguments and Its Fundamental Role in Nonmonotonic Reasoning, Logic Programming and *n*-Person Games. *Artificial Intelligence*, 77, 321–357.

Ennis, R. H. (1982). Identifying Implicit Assumptions. *Synthese*, 51, 61–86.

Fleming, J. (1961). Burdens of Proof. *Virginia Law Review*, 47, 51–70.

Freeman, J. B. (1988). *Thinking Logically*. Englewood Cliffs, N.J.: Prentice Hall.

Freeman, J. B. (1991). *Dialectics and the Macrostructure of Arguments: A Theory of Argument Structure*. Berlin: Walter de Gruyter.

Freeman, J. B. (1995). The Appeal to Popularity and Presumption by Common Knowledge. In *Fallacies: Classical and Contemporary Readings*, ed. H. V. Hansen and R. C. Pinto. University Park: Pennsylvania State University Press, 263–273.

Freeman, J. B. (2005). *Acceptable Premises*. Cambridge: Cambridge University Press.

Gigerenzer, G., Todd, P. M., and the ABC Research Group (1999). *Simple Heuristics That Make Us Smart*. Oxford: Oxford University Press.

Golden, H. L. (1994). Knowledge, Intent, System and Motive: A Much Needed Return to the Requirement of Independent Relevance. *Lousiana Law Review*, 55, 179–216.

Goodwin, J. (2010). How to Refute an Argument. Available at: http://www.public. iastate.edu/~goodwin/spcom322/refute.pdf, accessed November 26, 2010.

Gordon, T. F. (1995). *The Pleadings Game: An Artificial Intelligence Model of Procedural Justice*. Dordrecht: Kluwer.

Gordon, T. F. (2005). A Computational Model of Argument for Legal Reasoning Support Systems. In *Argumentation in Artificial Intelligence and Law*, IAAIL Workshop Series, ed. P. E. Dunne and T. Bench-Capon. Oisterwijk: Wolf Legal Publishers, 53–64.

Gordon, T. F. (2007). Visualizing Carneades Argument Graphs. *Law, Probability and Risk*, 6, 109–117.

Gordon, T. F. (2010). An Overview of the Carneades Argumentation Support System. In *Dialectics, Dialogue and Argumentation*, ed. C. Reed and C. W. Tindale. London: College Publications, 145–156.

Gordon, T. F., and Walton, D. (2006a). The Carneades Argumentation Framework. In *Computational Models of Argument: Proceedings of COMMA 2006*, ed. P. E. Dunne and T.J.M. Bench-Capon. Amsterdam: IOS Press, 195–207.

Gordon, T. F., and Walton, D. (2006b). Pierson v. Post Revisited. In *Computational Models of Argument: Proceedings of COMMA 2006*, ed. P. E. Dunne and T. J. M. Bench-Capon. Amsterdam: IOS Press, 208–219.

Gordon, T. F., and Walton, D. (2009). Proof Burdens and Standards. In *Argumentation in Artificial Intelligence*, ed. I. Rahwan and G. Simari. Berlin: Springer, 239–260.

Gordon, T. F., Prakken, H., and Walton, D. (2007). The Carneades Model of Argument and Burden of Proof. *Artificial Intelligence*, 171, 875–896.

Gough, J., and Tindale, C. (1985). Hidden or Missing Premises. *Informal Logic*, 7, 99–106.

Govier, T. (1992). *A Practical Study of Argument*, 3rd ed. Belmont, Calif.: Wadsworth.

Govier, T. (1999). *The Philosophy of Argument*. Newport News, Va.: Vale Press.

Govier, T. (2006). *The Philosophy of Argument*. Newport News, Va.: Vale Press.

Gray, B. E. (2002). Reported and Recommendations on the Law of Capture and First Possession: Popov v. Hayashi. *Superior of the State of California for the City and County of San Francisco*, Case no. 400545, November 6, 2002. Available at: http://web.mac.com/graybe/Site/Writings_files/Hayashi%20Brief.pdf, accessed May 24, 2009.

Grice, H. P. (1975). Logic and Conversation. In *The Logic of Grammar*, ed. D. Davidson and G. Harman. Encino, Calif.: Dickenson, 64–75.

Guarini, M. (2004). A Defense of Non-deductive Reconstructions of Analogical Arguments. *Informal Logic*, 24, 153–168.

Guarini, M., Butchart, A., Simard Smith, P., and Moldovan, A. (2009). Resources for Research on Analogy: A Multi-disciplinary Guide. *Informal Logic*, 29(2), 84–197.

Hamblin, C. L. (1970). *Fallacies*. London: Methuen.

Hamblin, C. L. (1971). Mathematical Models of Dialogue. *Theoria*, 37, 130–155.

Hamilton, W. (1861). *Discussions on Philosophy and Literature*. New York: Harper and Brothers.

Hamilton, W. (1874). *Lectures on Logic*. Edinburgh: William Blackwood and Sons.

Hansen, H. V., and Pinto, R. C. (eds.) (1995). *Fallacies: Classical and Contemporary Readings*. University Park: Pennsylvania State University Press, 251–264.

Hansen, H. V., and Walton, D. (2013). Argument Kinds and Argument Roles in the Ontario Provincial Election, *Journal of Argumentation in Context*, to appear.

Hart, H.L.A. (1949). The Ascription of Responsibility and Rights. In *Proceedings of the Aristotelian Society*, 49, 171–194. Reprinted in *Logic and Language*, ed. A. Flew (Oxford: Blackwell, 1951), 145–166.

Hart, H.L.A. (1961). *The Concept of Law*. Oxford: Oxford University Press.

Hitchcock, D. (1985). Enthymematic Arguments. *Informal Logic*, 7, 83–97.

Hitchcock, D., and Verheij, B. (eds.) (2006). *Arguing on the Toulmin Model: New Essays in Argument Analysis and Evaluation*. Dordrecht: Springer.

Hurley, P. (2003). *A Concise Introduction to Logic*, 8th ed. Belmont, Calif.: Wadsworth.

Irwin, T. (1988). *Aristotle's First Principles*. Oxford: Clarendon Press.

Jackson, S., and Jacobs, S. (1980). Structure of Conversational Argument: Pragmatic Bases for the Enthymeme. *Quarterly Journal of Speech*, 66, 251–165.

Jacovino, N. (1998). Red-Blooded Doctors Cure Anemia. *Harvard University Gazette*, January 22. Available at: http://www.news.harvard.edu/gazette/1998/01.22/Red-BloodedDoct.html, accessed February 3, 2009.

Johnson, R. H. (2000). *Manifest Rationality: A Pragmatic Theory of Argument.* Mahwah, N.J.: Erlbaum.

Johnson, R., and Blair, A. (1983). *Logical Self-Defence,* 2nd ed. Toronto: McGraw-Hill Ryerson.

Joseph, H.W.B. (1916). *An Introduction to Logic.* Oxford: Clarendon Press.

Josephson, J. R., and Josephson, S. G. (1994). *Abductive Inference: Computation, Philosophy, Technology.* New York: Cambridge University Press.

Krabbe, E.C.W. (1995). Appeal to Ignorance. In *Fallacies: Classical and Contemporary Readings,* ed. H. V. Hansen and R. C. Pinto. University Park: Pennsylvania State University Press, 251–264.

Krabbe, E.C.W. (1999). Profiles of Dialogue. In *JFAK: Essays Dedicated to Johan van Benthem on the Occasion of his 50th Birthday,* ed. J. Gerbrandy, M. Marx, M. de Rijke and Y. Venema. Amsterdam: University of Amsterdam Press, 25–36.

Krabbe, E.C.W. (2001). The Problem of Retraction in Critical Discussion. *Synthese,* 127, 141–159.

Krabbe, E.C.W. (2007). Nothing but Objections! In *Reason Reclaimed,* ed. H. V. Hansen and R. C. Pinto. Newport News, Va.: Vale Press, 51–63.

Krabbe, E.C.W. (2009). Review of Tindale (2007). *Argumentation,* 23, 127–131.

Kripke, S. (1965). Semantical Analysis of Intuitionistic Logic I. In *Formal Systems and Recursive Functions,* ed. J. N. Crossley and Michael Dummet. Amsterdam: North-Holland.

Leake, D. B. (1992). *Evaluating Explanations: A Content Theory.* Hillsdale, N.J.: Erlbaum.

Leonard, D. P. (2001). Character and Motive in Evidence Law. *Loyola of Los Angeles Law Review,* 34, 439–536.

Lewinski, M. (2008). The Paradox of Charity. *Informal Logic,* 32(4), 403–439.

Lewinski, M. (2011). Towards a Critique-friendly Approach to the Straw Man Fallacy Evaluation. *Argumentation,* 25(4), 469–497.

Loui, R. P. (1995). Hart's Critics on Defeasible Concepts and Ascriptivism. In *Proceedings of the Fifth International Conference on Artificial Intelligence and Law.* New York: ACM Press, 21–30. Available at: http://portal.acm.org/citation.cfm?id=222099.

Macagno, F., and Walton, D. (2009). Argument from Analogy in Law, the Classical Tradition, and Recent Theories. *Philosophy and Rhetoric,* 42, 154–182.

Mackenzie, J. D. (1981). The Dialectics of Logic. *Logique et Analyse,* 94, 159–177.

Mackenzie, J. D. (1990). Four Dialogue Systems. *Studia Logica,* 49, 567–583.

Magnani, L. (2001). *Abduction, Reason and Science.* New York: Kluwer.

McBurney, J. H. (1936). The Place of the Enthymeme in Rhetorical Theory. *Speech Monographs,* 3, 49–74.

McBurney, P., and Parsons, S. (2001a). Chance Discovery Using Dialectical Argumentation. In *New Frontiers in Artificial Intelligence,* ed. T. Terano, T. Nishida, A. Namatame, S. Tsumoto, Y. Ohsawa and T. Washio (Lecture Notes in Artificial Intelligence, vol. 2253). Berlin: Springer Verlag, 414–424.

McBurney, P., and Parsons, S. (2001b). Representing Epistemic Uncertainty by Means of Dialectical Argumentation. *Annals of Mathematics and Artificial Intelligence,* 32, 125–169.

McBurney, P., and Parsons, S. (2002). Games That Agents Play: A Formal Framework for Dialogues between Autonomous Agents. *Journal of Logic, Language and Information,* 11, 315–334.

McCarthy, J. (1986). Applications of Circumscription to Formalizing Common Sense Knowledge. *Artificial Intelligence*, 28, 89–116.

McCarthy, K. M. (2002). Statement of Decision. Superior Court of California, December 12, 2002, Case of Popov v. Hayashi #4005545: www.findlaw.

McCarty, L. T., and Sridharan, N. S. (1982). A Computational Theory of Legal Argument. LRP-TR-13. Laboratory for Computer Science Research. New Brunswick, N.J.: Rutgers University, 1–36.

McLaren, B. M. (2003). Extensionally Defining Principles and Cases in Ethics: An AI Model. *Artificial Intelligence Journal*, 150, 145–181.

McLaren, B. M. (2006). Computational Models of Ethical Reasoning: Challenges, Initial Steps, and Future Directions. In *IEEE Intelligent Systems*. Published by the IEEE Computer Society, July/August, 29–37.

Minot, G. R., and Murphy, W. P. (2001). Treatment of Pernicious Anemia by a Special Diet. *Yale Journal of Biology and Medicine*, 74, 341–353. Reprinted from the *Journal of the American Medical Association*, 87, 1926, 470–476.

Minsky, M. (1975). A Framework for Representing Knowledge. In *The Psychology of Computer Vision*, ed. P. Winston. McGraw-Hill. Available at: http://web.media.mit.edu/~minsky/papers/Frames/frames.html.

Misak, C. (1991). *Truth and the End of Inquiry: A Peircean Account of Truth*. Oxford: Clarendon Press.

Mochales, R., and Leven, A. (2009). Creating an Argument Corpus: Do Theories Apply to Real Arguments? In *Proceedings of the 12th International Conference on Artificial Intelligence and Law*. New York: Association for Computing Machinery, 21–30.

Mochales Palau, R., and Moens, M.-F. (2007). Study on Sentence Relations in the Automatic Detection of Argumentation in Legal Cases. In *Legal Knowledge and Information Systems: JURIX 2007, The Twentieth International Conference*, ed. A. Lodder and L. Mommers. Amsterdam: IOS Press, 89–98.

Mochales Palau, R., and Moens, M.-F. (2008). Study on the Structure of Argumentation in Case Law. In *Legal Knowledge and Information Systems: JURIX 2008, The Twenty-First International Conference*, ed. E. Francesconi, G. Sartor and D. Tiscornia. Amsterdam: IOS Press, 11–20, 89–98.

Mochales Palau, R., and Moens, M.-F. (2009). Argumentation Mining: The Detection, Classification and Structure of Arguments in Text. In *Proceedings of the 12th International Conference on Artificial Intelligence and Law*. New York: Association for Computing Machinery, 98–107.

Moens, M.-F., Mochales Palau, R., Boiy, E., and Reed, C. (2007). Automatic Detection of Arguments in Legal Texts. In *Proceedings of the International Conference on AI and Law* (ICAIL 2007), Stanford, Calif., 225–230.

Paglieri, F., and Woods, J. (2011). Enthymematic Parsimony. *Synthese*, 178, 461–501.

Pardo, M. S., and Allen, R. J. (2007). Juridical Proof and the Best Explanation. *Law and Philosophy* 27, 223–268.

Park, R. C., Leonard D. P., and Goldberg, S. H. (1998). *Evidence Law*. St. Paul, Minn.: West Group.

Patry, W. (2005/6). The Patry Copyright Blog. Available at: http://williampatry.blogspot.com/2005/06/striking-similarity-and-evidentiary.html, accessed July 22, 2010.

Peirce, C. S. (1931). *Collected Chapters*. Ed. C. Hartshorne and P. Weiss. Cambridge: Harvard University Press.

Peirce, C. S. (1984). *Writings of Charles S. Peirce: A Chronological Edition*, vol. 2. Ed. E. C. Moore. Bloomington: Indiana University Press.

Peirce, C. S. (1986). *Writings of Charles S. Peirce: A Chronological Edition*, vol. 3. Ed. Peirce Edition Project. Bloomington and Indianapolis: Indiana University Press.

Pennington, N., and Hastie, R. (1993). The Story Model for Juror Decision Making. In *Inside the Juror: The Psychology of Juror Decision Making*, ed. R. Hastie. Cambridge: Cambridge University Press, 192–221.

Perelman, C., and Olbrechts-Tyteca, L. (1969). *The New Rhetoric*. Notre Dame, Ind.: University of Notre Dame Press.

Pollock, J. (1995). *Cognitive Carpentry*. Cambridge, Mass.: MIT Press.

Popper, K. (1963). *Conjectures and Refutations: The Growth of Scientific Knowledge*. Routledge: London.

Popper, K. (1972). *Objective Knowledge: An Evolutionary Approach*. Oxford: Oxford University Press, 1972.

Prakken, H. (2000). On Dialogue Systems with Speech Acts, Arguments and Counterarguments. In *Proceedings of JELIA 2000, the European Workshop on Logic for Artificial Intelligence*, ed. M. Ojeda-Aciego, I. P. de Guzman, G. Brewka and L. M. Pereira. Berlin: Springer, 224–238.

Prakken, H. (2003). Logical Dialectics: The Missing Link between Deductivism and Pragma-Dialectics. In *Proceedings of the Fifth Conference of the International Society for the Study of Argumentation*, ed. Frans H. van Eemeren et al. Amsterdam: SicSat, 857–860.

Prakken, H. (2005). Coherence and Flexibility in Dialogue Games for Argumentation. *Journal of Logic and Computation*, 15, 1009–1040.

Prakken, H. (2006). Formal Systems for Persuasion Dialogue. *Knowledge Engineering Review*, 21, 163–188.

Prakken, H. (2010). On the Nature of Argument Scheme. In *Dialectics, Dialogue and Argumentation*, ed. C. Reed and C. W. Tindale. London: College Publications, 167–185.

Prakken, H., and Sartor, G. (1996). A Dialectical Model of Assessing Conflicting Arguments in Legal Reasoning. *Artificial Intelligence and Law*, 4, 331–368.

Prakken, H., and Sartor, G. (1997). Argument-Based Extended Logic Programming with Defeasible Priorities. *Journal of Applied Non-classical Logics*, 7, 25–75.

Prakken, H., and Sartor, G. (2006a). A Dialectical Model of Assessing Conflicting Arguments in Legal Reasoning. *Artificial Intelligence and Law*, 4, 331–368.

Prakken, H., and Sartor, G. (2006b). Presumptions and Burdens of Proof. In *Legal Knowledge and Information Systems: JURIX 2006: The Nineteenth Annual Conference*, ed. T. M. van Engers. Amsterdam: IOS Press, 21–30.

Prakken, H., and Sartor, G. (2007). Formalising Arguments about the Burden of Persuasion. In *Proceedings of the Eleventh International Conference on Artificial Intelligence and Law*. New York: ACM Press, 97–106.

Prakken, H., and Sartor, G. (2009). A Logical Analysis of Burdens of Proof. In *Legal Evidence and Proof: Statistics, Stories, Logic*, ed. H. Kaptein, H. Prakken and B. Verheij. Farnham: Ashgate Publishing, 223–253.

Rahwan, I., Banihashemi, B., Reed, C., Walton, D., and Abdallah, S. (2011). Representing and Classifying Arguments on the Semantic Web. *Knowledge Engineering Review*, 26(4), 487–511.

Reed, C. (1998). Dialogue Frames in Agent Communication. In *Proceedings of the Third International Conference on Multi-Agent Systems*, ed. Y. Demazeau. IEEE Press, 246–253.

Reed, C. (2006). Representing Dialogic Argumentation. *Knowledge-Based Systems*, 19(1), 22–31.

Reed C., and Grasso, F. (2007). Recent Advances in Computational Models of Natural Argument. *International Journal of Intelligent Systems*, 22(1), 1–15.

Reed, C., and Walton, D. (2003). Diagramming, Argumentation Schemes and Critical Questions. In *Anyone Who Has a View: Theoretical Contributions to the Study of Argumentation*, ed. F. H. van Eemeren, J. A. Blair, C. A. Willard and A. Snoek Henkemans. Dordrecht: Kluwer, 195–211.

Reed, C., and Tindale, C. W. (eds.) (2010). *Dialectics, Dialogue and Argumentation: An Examination of Douglas Walton's Theories of Reasoning and Arguments*. London: College Publications.

Reed, C., Walton, D., and Macagno, F. (2007). Argument Diagramming in Logic, Law and Artificial Intelligence. *Knowledge Engineering Review*, 22, 87–109.

Reiter, R. (1980). A Logic for Default Reasoning. *Artificial Intelligence*, 13, 81–132.

Reiter, R. (1987). Nonmonotonic Reasoning. *Annual Review of Computer Science*, 2, 147–186.

Rescher, N. (1977). *Dialectics: A Controversy-Oriented Approach to the Theory of Knowledge*. Albany: State University of New York Press, 1977.

Restificar, A., Ali, S., and McRoy, S. (1999). ARGUER: Using Argument Schemas for Argument Detection and Rebuttal in Dialogs. In *UMP99: International Conference on User Modeling*, ed. Judy Kay. New York: Springer-Wien, 315–317.

Ribeiro, B. (2008). How Often Do We (Philosophy Professors) Commit the Straw Man Fallacy? *Teaching Philosophy*, 31, 27–38.

Rissland, E., and Ashley, K. (1987). A Case-Based System for Trade Secrets Law. In *Proceedings of the First International Conference on Evidence and Law*, Boston, 60–66.

Robinson, R. (1953). *Plato's Earlier Dialectic*. Oxford: Clarendon Press.

Robinson, R. (1962). *Plato's Earlier Dialectic*, 2nd ed. Oxford: Clarendon Press.

Ross, W. D., ed. (1928). *The Works of Aristotle Translated into English*, vol. 1. Trans. W. A. Pickard-Cambridge. Oxford: Oxford University Press.

Russell, S., and Norvig, P. (1995). *Artificial Intelligence: A Modern Approach*. Upper Saddle River, N.J.: Prentice-Hall.

Salmon, W. (1963). *Logic*. Englewood Cliffs, N.J.: Prentice-Hall.

Sartor, G. (2005). *Legal Reasoning: A Cognitive Approach to the Law*. Springer: Berlin.

Schank, R. C. (1986). *Explanation Patterns: Understanding Mechanically and Creatively*. Hillsdale, N.J.: Erlbaum.

Schank, R. C., and Abelson, R. P. (1977). *Scripts, Plans, Goals and Understanding*. Hillsdale, N.J.: Erlbaum.

Schauer, F. (1987). Precedent. *Stanford Law Review*, 39(3), 571–605.

Schauer, F. (2009). *Thinking like a Lawyer*. Cambridge, Mass.: Harvard University Press.

Scheuer, O., Loll, F., Pinkwart, N., and McLaren, B. M. (2010). Computer-Supported Argumentation: A Review of the State of the Art. *Computer-Supported Collaborative Learning*, 5(1), 43–102.

Scriven, M. (1976). *Reasoning*. New York: McGraw-Hill.

Searle, J. (2001). *Rationality in Action.* Cambridge, Mass.: MIT Press.

Sergot, M., Sadri, A., Kowalski, R., Kriwaczek, F., Hammond P., and Cory, H. T. (1986). The British Nationality Act as a Logic Program. In *Proceedings ZNP-83 Congress,* ed. G. Van Nevel and F. Balfroid. New York: Elsevier North-Holland, 29(5), 370–386.

Singh, M. P. (2000). A Social Semantics for Agent Communication Languages. In *Issues in Agent Communication,* ed. F. Dignum and M. Greaves. Berlin: Springer-Verlag, 31–45.

Singh, P., Lin, T., Mueller, E., Lim, G., Perkins, T., and Li Zhu, W. (2002). Open Mind Common Sense: Knowledge Acquisition from the General Public. In *Proceedings of the First International Conference on Ontologies, Databases, and Applications of Semantics for Large Scale Information Systems,* Lecture Notes in Computer Science. Heidelberg: Springer, 1123-1237.

Tamminga, A. (2001). Belief Dynamics. Ph.D. thesis. Institute for Logic, Language and Computation, University of Amsterdam.

Thomson, J. (1971). A Defense of Abortion. *Philosophy and Public Affairs,* 1(1), 47–66.

Tillers, P., and Gottfried J. (2006). Case comment – *United States v. Copeland,* 369 F. Supp. 2d 275 (E.D.N.Y. 2005): A Collateral Attack on the Legal Maxim That Proof Beyond a Reasonable Doubt Is Unquantifiable? *Law, Probability and Risk,* 5, 135–157.

Tindale, C. W. (1997). Fallacies, Blunders and Dialogue Shifts: Walton's Contributions to the Fallacy Debate. *Argumentation,* 11, 341–354.

Tindale, C. W. (1999). *Acts of Arguing: A Rhetorical Model of Argument.* Albany: State University of New York Press.

Tindale, C. (2007). *Fallacies and Argument Appraisal.* Cambridge: Cambridge University Press.

Toulmin, S. (1958). *The Uses of Argument.* Cambridge: Cambridge University Press.

Van Eemeren, F. H. (2010). *Strategic Maneuvering in Argumentative Discourse.* Amsterdam: Benjamins.

Van Eemeren, F. H. and Grootendorst, R. (1984). *Speech Acts in Communicative Discussions.* Dordrecht: Foris.

Van Eemeren, F. H., and Grootendorst, R. (1987). Fallacies in Pragma-Dialectical Perspective. *Argumentation,* 1, 283–301.

Van Eemeren, F. H., and Grootendorst, R. (1992). *Argumentation, Communication and Fallacies.* Mahwah, N.J.: Erlbaum.

Van Eemeren, F. H., and Houtlosser, P. (2006). Strategic Maneuvering: A Synthetic Recapitulation. *Argumentation,* 20, 381–392.

Van Eemeren, F. H., Houtlosser, P., and Snoeck Henkemans, F. (2007). *Argumentative Indicators in Discourse.* Dordrecht: Springer.

Verheij, B. (2003). Dialectical Argumentation with Argumentation Schemes: An Approach to Legal Logic. *Artificial Intelligence and Law,* 11, 167–195.

Verheij, B. (2005). *Virtual Arguments. On the Design of Argument Assistants for Lawyers and Other Arguers.* The Hague: TMC Asser Press.

Verheij, B. (2009). The Toulmin Argument Model in Artificial Intelligence. In *Argumentation in Artificial Intelligence,* ed. I. Rahwan and G. Simari. Berlin: Springer, 219–238.

Vreeswijk, G., and Prakken, H. (2000). Credulous and Sceptical Argument Games for Preferred Semantics. In *Proceedings of the 7th European Workshop on Logics in Artificial Intelligence*, Springer Lecture Notes in AI 1919, Berlin: Springer Verlag, 239–253.

Wagenaar, W. A., van Koppen, P. J., and Crombag, H.F.M. (1993). *Anchored Narratives: The Psychology of Criminal Evidence*. Hertfordshire: Harvester Wheatsheaf.

Walton, D. (1984). *Logical Dialogue-Games and Fallacies*. Lanham, Md.: University Press of America.

Walton, D. (1990). *Practical Reasoning: Goal-Driven, Knowledge-Based, Action-Guiding Argumentation*. Savage, Md.: Rowman & Littlefield.

Walton, D. (1992). *Slippery Slope Arguments*. Oxford: Oxford University Press.

Walton, D. (1995). *A Pragmatic Theory of Fallacy*. Tuscaloosa: University of Alabama Press, 1995.

Walton, D. (1996a). *Argument Structure: A Pragmatic Theory*. Toronto: University of Toronto Press.

Walton, D. (1996b). *Argumentation Schemes for Presumptive Reasoning*. Mahwah, N.J.: Erlbaum.

Walton, D. (1996c). *Arguments from Ignorance*. University Park: Penn State University Press.

Walton, D. (1996d). The Straw Man Fallacy. In *Logic and Argumentation*, ed. J. van Benthem, F. H. van Eemeren, R. Grootendorst and F. Veltman. Amsterdam: North-Holland, 115–128.

Walton, D. (1997). *Appeal to Expert Opinion*, University Park: Penn State University Press.

Walton, D. (1998). *The New Dialectic: Conversational Contexts of Argument*. Toronto: University of Toronto Press.

Walton, D. (1999a). Profiles of Dialogue for Arguments from Ignorance. *Argumentation*, 13, 53–71.

Walton, D. (1999b). Rethinking the Fallacy of Hasty Generalization. *Argumentation*, 13, 161–182.

Walton, D. (2001). Enthymemes, Common Knowledge and Plausible Inference. *Philosophy and Rhetoric*, 34, 93–112.

Walton, D. (2002). The Sunk Costs Fallacy or Argument from Waste. *Argumentation*, 16, 473–503.

Walton, D. (2003). Is There a Burden of Questioning? *Artificial Intelligence and Law*, 11, 1–43.

Walton, D. (2006a). Argument from Appearance: A New Argumentation Scheme. *Logique et Analyse*, 195, 2006, 319–340.

Walton, D. (2006b). *Character Evidence: An Abductive Theory*. Berlin: Springer.

Walton, D. (2006c). *Fundamentals of Critical Argumentation*. New York: Cambridge University Press.

Walton, D. (2006d). Poisoning the Well. *Argumentation*, 20, 273–307.

Walton, D. (2007a). *Dialog Theory for Critical Argumentation*. Amsterdam: John Benjamins Publishing.

Walton, D. (2007b). Metadialogues for Resolving Burden of Proof Disputes. *Argumentation*, 21, 291–316.

Walton, D. (2008a). *Informal Logic*, 2nd ed. Cambridge: Cambridge University Press, 2008.

Walton, D. (2008b). Proleptic Argumentation. *Argumentation & Advocacy*, 44, 143–154.

Walton, D. (2008c). The Three Bases for the Enthymeme: A Dialogical Theory. *Journal of Applied Logic*, 6, 361–379.

Walton, D. (2008d). *Witness Testimony Evidence*. Cambridge: Cambridge University Press.

Walton, D. (2010a). A Dialogue Model of Belief. *Argument and Computation*, 1, 23–46.

Walton, D. (2010b). Why Fallacies Appear to Be Better Arguments Than They Are. *Informal Logic*, 30(2), 159–184.

Walton, D. (2011a). A Dialogue System Specification for Explanation. *Synthese*, 182(3), 349–374.

Walton, D. (2011b). Reasoning about Knowledge Using Defeasible Logic. *Argument and Computation*, 2(2–3), 131–155.

Walton, D. (2012). Using Argumentation Schemes for Argument Extraction: A Bottom-Up Method. *International Journal of Cognitive Informatics and Cognitive Computing*, 6(3), 33–60.

Walton, D., and Godden, D. M. (2005). The Nature and Status of Critical Questions in Argumentation Schemes. In *The Uses of Argument: Proceedings of a Conference at McMaster University*, ed. D. Hitchcock. Hamilton, Ontario: OSSA, 476–484.

Walton, D., and Gordon, T. F. (2005). Critical Questions in Computational Models of Legal Argument. In *Argumentation in Artificial Intelligence and Law*, IAAIL Workshop Series, ed. P. E. Dunne and T.J.M. Bench-Capon. Nijmegen: Wolf Legal Publishers, 103–111.

Walton, D., and Gordon, T. F. (2009). Jumping to a Conclusion: Fallacies and Standards of Proof. *Informal Logic*, 29, 215–243.

Walton, D., and Krabbe, E. C. W. (1995). *Commitment in Dialogue*. Albany: State University of New York Press.

Walton, D., and Macagno, F. (2010). Wrenching from Context: The Manipulation of Commitments. *Argumentation* 24(3), 283–317.

Walton, D., and Macagno, F. (2011). Quotations and Presumptions: Dialogical Effects of Misquotations. *Informal Logic*, 31(1), 26–54.

Walton, D., and Reed, C. (2005). Argumentation Schemes and Enthymemes. *Synthese: An International Journal for Epistemology, Methodology and Philosophy of Science*, 145, 339–370.

Walton, D., and Schafer, B. (2006). Arthur, George and the Mystery of the Missing Motive: Towards a Theory of Evidentiary Reasoning about Motives. *International Commentary on Evidence*, 4(2), 1–47.

Walton, D., Reed, C., and Macagno, F. (2008). *Argumentation Schemes*. Cambridge: Cambridge University Press.

Weinreb, L. L. (2005). *Legal Reason: The Use of Analogy in Legal Argument*. Cambridge: Cambridge University Press.

Wigmore, J. H. (1931). *The Principles of Judicial Proof*, 2nd ed. Boston: Little, Brown.

Wigmore, J. H. (1940). *Evidence in Trials at Common Law*. Boston: Little, Brown.

Williams, A. R. (2003). Burdens and Standards in Civil Litigation. *Sydney Law Review*, 25, 165–188.

Williams, G. (1977). The Evidential Burden: Some Common Misapprehensions. *New Law Journal*, Feb. 17, 156–158.

Wooldridge, M. (2000). *Reasoning about Rational Agents.* Cambridge, Mass.: MIT Press.

Wooldridge, M., and Jennings, N. R. (1995). Intelligent Agents: Theory and Practice. *Knowledge Engineering Review,* 10, 115–152.

Wyner, A., and Bench-Capon, T.J.M. (2007). Argument Schemes for Legal Case-Based Reasoning. In *Legal Knowledge and Information Systems: JURIX 2007, The Twentieth International Conference,* ed. A. Lodder and L. Mommers. Amsterdam: IOS Press, 139–149.

Wyner, A., Bench-Capon, T.J.M., and Atkinson, K. (2007). Arguments, Values and Baseballs: Representation of Popov v. Hayashi. In *Legal Knowledge and Information Systems* (JURIX 2007), ed. A. Lodder and L. Mommers. Amsterdam: IOS Press, 151–160.

Index

Printed in the United States
By Bookmasters